Network Technology for Digital Audio

Music
TECHNOLOGY
Series

Titles in the Series

Acoustics and Psychoacoustics, 2nd edition
(with accompanying website :
http://www-users.york.ac.uk/~dmh8/AcPsych/acpsyc.htm)
David M. Howard and James Angus

The Audio Workstation Handbook
Francis Rumsey

Composing Music with Computers (with CD-ROM)
Eduardo Reck Miranda

Computer Sound Synthesis for the Electronic Musician
(with CD-ROM)
Eduardo Reck Miranda

Digital Audio CD and Resource Pack
Markus Erne
(Digital Audio CD also available separately)

Network Technology for Digital Audio
Andy Bailey

Digital Sound Processing for Music and Multimedia
(with accompanying website:
http://www.York.ac.uk/inst/mustech/dspmm.htm)
Ross Kirk and Andy Hunt

MIDI Systems and Control, 2nd edition
Francis Rumsey

Sound and Recording: An Introduction, 3rd edition
Francis Rumsey and Tim McCormick

Sound Synthesis and Sampling
Martin Russ

Sound Synthesis and Sampling CD-ROM
Martin Russ

Network Technology for Digital Audio

Andy Bailey

Focal Press

OXFORD   AUCKLAND   BOSTON   JOHANNESBURG   MELBOURNE   NEW DELHI

Focal Press
An imprint of Butterworth-Heinemann
Linacre House, Jordan Hill, Oxford OX2 8DP
225 Wildwood Avenue, Woburn, MA 01801-2041
A division of Reed Educational and Professional Publishing Ltd

A member of the Reed Elsevier plc group

First published 2001

British Library Cataloguing in Publication Data
Bailey, Andy
    Network technology for digital audio
    1. Sound – Recording and reproducing – Digital techniques –
    Computer networks
    I. Title
    621.3893'028546

Library of Congress Cataloguing in Publication Data
Bailey, Andy.
    Network technology for digital audio/Andy Bailey.
    p. cm. – (Music technology series)
    Includes bibliographical references and index.
    ISBN 0-240-51588-9 (alk. paper)
    1. Digital communications.   2. RDS (Radio)
    3. Computer networks.   I. Title   II. Series.
    TK5103.7.B355
    621.382'1–dc21   00–052125

ISBN 0 240 51588 9

Composition by Genesis Typesetting, Laser Quay, Rochester, Kent
Printed and bound in Great Britain

FOR EVERY TITLE THAT WE PUBLISH, BUTTERWORTH-HEINEMANN
WILL PAY FOR BTCV TO PLANT AND CARE FOR A TREE.

Contents

Series introduction

The Focal Press Music Technology Series is intended to fill a growing need for authoritative books to support college and university courses in music technology, sound recording, multimedia and their related fields. The books will also be of value to professionals already working in these areas and who want either to update their knowledge or to familiarize themselves with topics that have not been part of their mainstream occupations.

Information technology and digital systems are now widely used in the production of sound and the composition of music for a wide range of end uses. Those working in these fields need to understand the principles of sound, musical acoustics, sound synthesis, digital audio, video and computer systems. This is a tall order, but people with this breadth of knowledge are increasingly sought after by employers. The series will explain the technology and techniques in a manner which is both readable and factually concise, avoiding the chattiness, informality and technical woolliness of many books on music technology. The authors are all experts in their fields and many come from teaching and research backgrounds.

Dr Francis Rumsey
Series Editor

Preface

Books of a technical nature suffer from a number of problems. Perhaps the most obvious is that technology progresses. What was once considered best practice in a particular field, may be regarded with derision at a later stage. Furthermore, recognized practices were once not recognized, whilst a book remains as a permanent record.

This book is about the transfer of audio data and other audio related data, over digital communications networks, and it is published at an exciting time for this specialist area. Although the Internet has made networks the property of the people, there were not previously the data rates available at the right cost to transfer the quantities of data (quantities that seemed impossible ten years ago, but will be shrugged off in half that time) required for professional audio use.

At the same time that data rates improved through advancements in the communications industry, the audio industry also looked for flexibility in transferring digital audio and control data. The meeting of technologies has produced a number of initiatives, specifications, and standard agreements aimed at audio applications. At the time of writing, no clear industry standard had emerged, although the collective works are available across a range of audio applications.

This book presents as many of the candidates as possible, with the provision that new initiatives emerge regularly and that current work is constantly updated, amended, improved or otherwise changed – to reflect the latest technology progression or recognized practise. To this end, further reading sections are supplied to help the reader seek out the most current information. Where possible, Internet addresses are supplied to help speed the process, although since these are subject to change, the postal address is also provided. On occasion, subject matter is included only to expose its presence, and since technology progresses, these may or may not be included in any audio future.

Another of the aforementioned problems of technical books is the tendency towards jargon, especially acronyms. Although no apologies are made for the use of acronyms, as to repeat the words verbatim at each appearance in the text would win praise from no one, sympathy is offered instead!

In defence, one of the goals of this book is to introduce digital communications to those already familiar with analogue theory. It is necessary therefore to create a foundation of concepts, terminology, and methods upon which to build knowledge. This is especially so since digital and analogue communications techniques, whilst apparently performing the same task, operate in a completely different fashion.

The first of the concepts that is introduced is the modular approach of the International Standards Organization open systems interconnection 7-layer reference model. This is simply a conceptual model designed to communicate and illustrate the steps required to achieve useful interdevice communications between two points, over a complex path (or network).

A useful by-product of including the 7-layer model is the useful introduction to conceptual models as a whole. Several of the technologies presented herein use their own established conceptual model to illustrate their methods. So-called 'technology evangelists' will insist that a particular technology does not fit into the 7-layer model and so a new conceptual model was created. On the whole, such a claim is misrepresentative since usually it is the case that the new conceptual model serves to focus in on a particular area of functionality within the 7-layer model, expanding and detailing that area. This is because the 7-layer model is such a general overview of areas requiring attention, whilst the conceptual models associated with particular technologies offer detailed descriptions of particular implementations.

Finally, it was a struggle to visualize who would find use for a book that spans the data communication and audio engineering fields. As it happens, this is now a very fast moving technological field, although it this has been an interest of mine since the early days of my audio training, as the following narrative, kindly supplied by my good friend Robert C. Alexander, illustrates:

Andy Bailey and I had been living and working in Australia for nearly a year when walking back from the recording studio in October 1990 one night, our discussion turned to the possibilities of the future of the audio industry.

These were the days when digital data transfer, ISDN, and mass information communication used as tools for the audio industry were in their infancy. Even the Internet was yet to make an impact. Just months earlier, the first murmuring of digital audio transfers between studios were being heralded. Indeed, digital recording itself was still a novelty.

We discussed the exchange of massive amounts of information between one point and another, allowing audio data and even video information to be exchanged freely and easily to anybody and everybody connected to a huge central communication system. No longer would artists need to fly at great expense all over the world to record albums in Los Angeles, London, New York, or Sydney. It could be done from the local recording facility using new communications technology. Entire mixes could be done from remote locations utilizing the engineers and producers most suited to the job, yet without them ever having to leave the facility in which they most preferred to work.

This represented a massive (and at that time unattainable) leap forward for the audio and computer industries, which had, as yet, to wake up to ideas such as the Internet. There were no publicly available communications systems that were fast enough to handle the amounts of data we were imagining; whilst user interfaces which could be accessed by everybody from audio engineers to laymen were over four years into the future. In hindsight, an amazing two hours saw us cover topics that we realized could happen, if they could be imagined by us.

In the years that followed, after returning to England, Andy pursued a different branch of the industry to myself, concentrating on technical education whilst I concerned myself with audio engineering, education, and writing. I had no hesitation in getting Andy involved with writing and I was justly rewarded with a continual flowing stream of articles that first brought the name Andrew Bailey to industry awareness. It came as no surprise to me when he was commissioned to write this much needed reference book for the audio industry.

I have no doubt that this complex subject will become one of the key elements in bringing the communication and information age that the twenty-first century promises, to the audio world.

This book is an extension of the conversation Andy and I had, a decade ago, walking home in Sydney.

Andy Bailey

Acknowledgements

Naturally, this book is dedicated to the long-suffering Hilary without whom I could not have completed it – a book is a serious undertaking at any time, never mind experiencing a family for the first time as well! Special thanks are reserved for Margaret Riley, Beth Howard, Jennifer Welham and all at Focal Press for their patience on so many missed deadlines. Additional thanks to Francis Rumsey, who has been a gentle and patient teacher and to Rob Alexander for his encouragement, friendship, and experience.

Matt Roberts produced excellent work on all the diagrams whilst finishing a degree. Colin Tugwood and Kenton White found reference material for me, as did Tony Ruane at Redstor. Other help was also provided by Eddie Kenehan, Michael Krueger, Jens Thaler, Phil Welsh, Andy Cheshire, the management and my colleagues at gedas.

Thanks also to the BBC, the AES, and John Watkinson, along with all those who have been sourced. Jayson Chase and Nathan also deserve a mention for their additional effort.

1 Digital audio and computer networks

The subject of digital audio transfer, by necessity, draws heavily from two technological industries. The first is the audio industry and the second is the computer networking industry. The marrying of these two industries has been part of the greater trend towards a larger multimedia industry, fuelled by the accessibility of computer technology as well as audio and visual quality to the consumer.

The first chapter in this book is intended for those readers without a background in one or other of these subjects and introduces some of the terminology and fundamental concepts required to understand the arguments for and against the candidate solutions that are presented. The two subjects are regularly placed within the context of the counterpart, since technology in one area can be described and understood by looking at the theory developed for the other.

1.1 What is digital?

The term digital means a signalling scheme in which data are represented as numbers. Most commonly, but not always, data are represented as one of two states. From such a vague statement, almost anything can represent two states. For example, 1 and 0, dot and dash, or north and south. Provided each state is always represented in the same way it only really matters that they are always interpreted in the same way.

The normal every day numbering system is based on ten numbers (0 to 9) and is called decimal, or base-10. The mathematics surrounding the numbering system based on only two numbers is called binary or base-2.

Binary numbering systems use 1 and 0 as the two states, and this translates electrically so that 1 represents on and 0 represents off. It is also perfectly acceptable for 1 to represent off and 0 to represent on, but the former assignments are used most commonly.

Once binary information has been represented in this way, it is possible to use a series of switches to perform complex mathematical operations upon the numbers, by operating the switches according to the rules laid out by the mathematical, or logical, operation.

1.1.1 Binary and hex

Any number can be represented in binary, since if a number is equal to more than 1, then it is carried forward, in the same way as elementary decimal addition for the decimal numbering system. Obviously, the amount of times the carrying forward occurs is likely to be higher, since writing down the number 2 would be the first whole number that requires this to happen, and would be represented as 1 0 (see Figure 1.1).

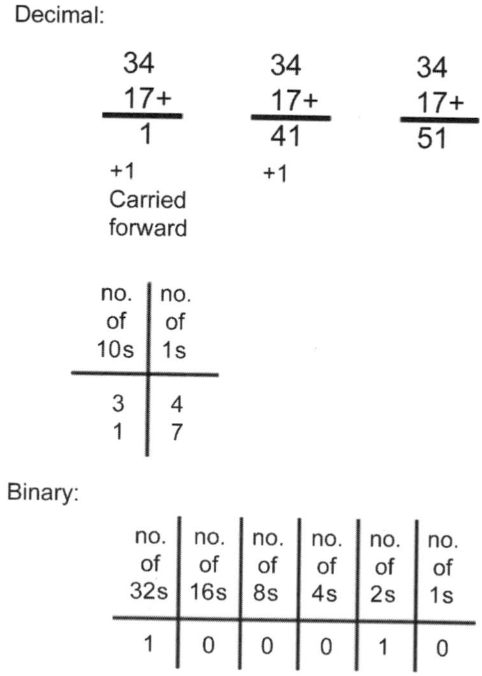

Decimal:

$$
\begin{array}{ccc}
34 & 34 & 34 \\
17+ & 17+ & 17+ \\
\hline
1 & 41 & 51 \\
\end{array}
$$

+1
Carried
forward

+1

no. of 10s	no. of 1s
3	4
1	7

Binary:

no. of 32s	no. of 16s	no. of 8s	no. of 4s	no. of 2s	no. of 1s
1	0	0	0	1	0

Figure 1.1 Binary mathematics. In everyday decimal mathematics, numbers are represented as rows of numbers, each row representing a power of the base number (1, 10, 100, 1000) and so on. Binary mathematics can also be represented as columns of numbers, incrementing by a power of 2 (1, 2, 4, 8, 16 and so on).

By using electrical voltages and switches, computing devices have been able to make valuable use of binary representation of numbers. All computer data in its most basic form are made up of binary digits. As might be expected from the impressive rate of development of computer technology, the binary method of representing information comes with a complete set of jargon.

In binary digital devices, the most common methods for representing bi-states include voltage differences, magnetic polarity, or the intensity of light.

A single bi-state position (either a 1 or a 0) is called a bit, from BInary digiTs. Bits are grouped together to form 'words'. It is common practice in computing to group 8 bits together to form an 8-bit word length. An 8-bit word is called a byte and is something that was settled on through trial-and-error and due to the limitation of early computer designs, and has been around for over fifty years.

Eight bits together can represent any number up to 256 ($2 \times 2 \times 2 \times 2 \times 2 \times 2 \times 2 \times 2$, written as 2^8 for ease).

To make bytes useful for carrying information, various methods have been devised for splitting them up in different ways to represent different things.

Figure 1.2 shows two common bit assignments. The first is into two equal halves, each half being used for a different function. For instance, the leftmost half of the byte may be used to represent a communications channel or address, which is used to identify another device on a network. In this example, the rightmost half represents a function to be performed. In this way, devices at a specific address may perform a certain function, while other devices with different addresses will ignore this instruction. Using this example, it would then be possible to

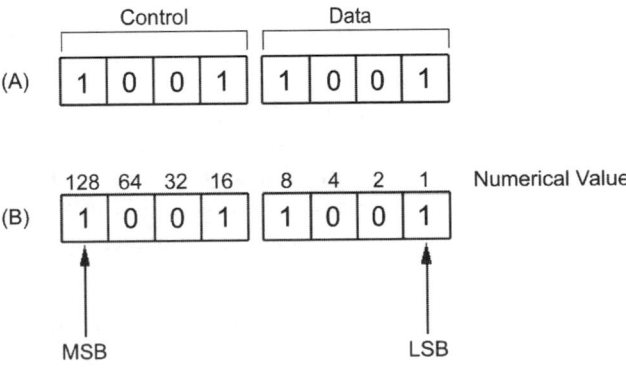

Figure 1.2 Common bit assignments. (A) Control data and payload. (B) Most significant bit (MSB) and least significant bit (LSB). Bits can be both at the same time.

build up a control set for simple devices such as might be used to control a toy car or robotic arm. This split of an 8-byte word into two halves became so common that 4 bits got its own jargon and is so called a nibble (the amusing principle being that a nibble is half a byte).

A second common way of looking at bytes, also shown in Figure 1.2, is to assign an importance to each position and this is not an exclusive way of considering bits within a byte. This is easy to understand, since all the bits might appear equal in importance if no other information regarding the information the bits represent is available. The only exceptions are the two bits at either end of a transmission or string of bits, if only because they have only one neighbour and the others have two. As a result these are given different importance and assigned the titles of most significant bit (MSB) and least significant bit (LSB). When written down, the MSB is the leftmost bit, and in mathematical terms has the greatest weight ($2^{(n-1)}$ where n is the number of bits).

Since this 8-bit word length is so popular, the shorthand for representing digital data is to use hexadecimal annotation. Hexadecimal or *hex*, is a base-16 numbering system, meaning that 16 numbers can be represented before a carry forward needs to be performed. It is convenient to annotate 8-bit bytes using such a numbering system, because 4 bits can represent 16. If the numbering system can represent 16 before the first carry over, then the 8-bit byte can be written in two positions.

It is difficult to represent 16 unique digits without carrying over because we are so used to base-10. The convention is to use letters following the number 9. Therefore, hex uses the numbers 0, 1, 2, 3, 4, 5, 6, 7, 8, 9, and letters A, B, C, D, E, F. Using this system, an 8-bit word would be represented using an annotation such as FF or 9D (255 and 157, respectively, in decimal format). Confusion can arise when a hexadecimal number contains only the numbers, rather than with letters, and so an indicator is often placed before or after the number such as x88, 99(h) or h99. This book will use the format 99h.

As time progressed, 8-bit computing was replaced by more powerful multiplications, such as 16- 32- 64- and 128-bit manipulation (and so on), but hexadecimal remains a powerful annotation tool, and is still used when addressing large numbers of devices on computer networks such as the Internet.

As the amount of data capable of being manipulated, stored and moved increased, it became common to use Latin terms to represent the order of magnitude. It should be noted that within computer technology, the order of magnitude is intrinsically related to the binary numbering system. As a result, round numbers are those that are an exact factor of 2. So for instance, the multiples of interest are 2, 4, 8, 16, and so on, reflecting the bit position value in a word. Larger factors are given useful names, so a kilobyte = 1024 bytes, megabyte = 1024 kilobytes, a gigabyte = 1024 megabytes, and a terabyte = 1024 Gigabytes and so on.

It should be noted that the counting system used in binary bases itself on a doubling at each order of magnitude, demonstrating a similar behaviour to frequency.

1.2 What is digital audio?

By using huge numbers of bits it is possible to represent a great deal of information, and once an understanding of what each bit means has been established, it is possible to write computer programs to manipulate the information, providing an understanding of the meaning of each bit also ensures that the mechanism and structures used for representing data are understood. There are a number of different mechanisms used throughout the audio industry, representing different types of data. Data types include control data, timing, and synchronization information as well as audio signals transferred to binary representation. Each data type has its own storage and transmission requirements, data structure, and rules for interpretation.

1.2.1 Audio data

Among the short list of data types above, probably the most significant type, at least in terms of size, is the representation of an analogue audio signal as digital data. Analogue audio waves are converted into binary information using a method known as pulse code modulation (PCM). The ear can only interpret analogue audio information, so despite the benefits associated with manipulating audio in a digital environment, the signal must be turned back into an analogue waveform in order to be heard. In analogue technology a sound wave is recorded or used in its analogue form. In the simple analogue tape recording example shown in Figure 1.3, a signal is taken from the microphone, amplified to a suitable level for the tape machine, and then stored on magnetic tape. The sound waves from the source to the microphone, from the microphone to the amplifier and from the amplifier to the recording machine are all analogue waves, as is the information on the tape itself. The tape can be

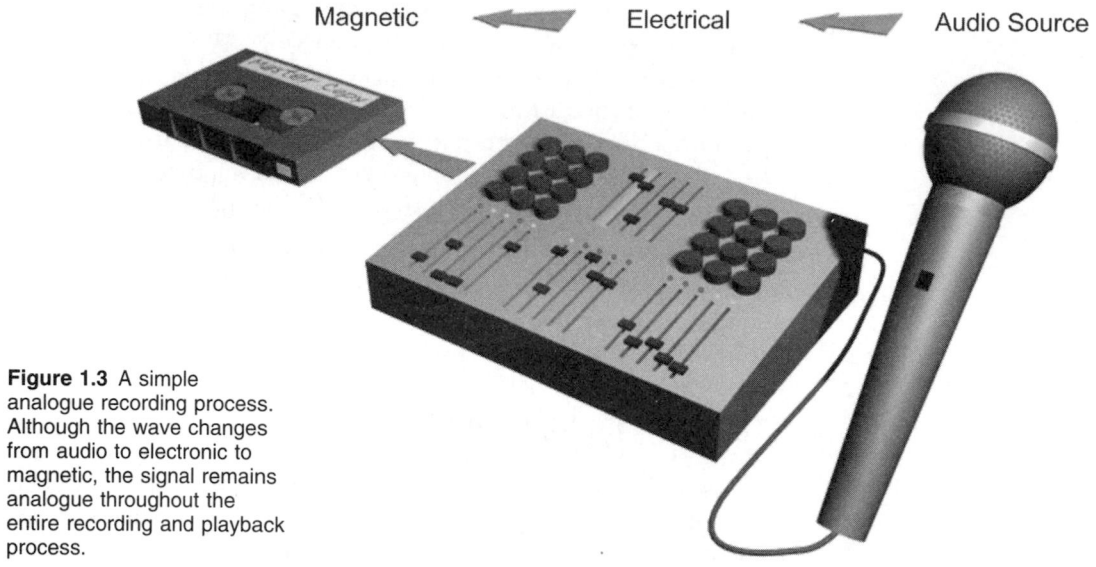

Magnetic ← Electrical ← Audio Source

Figure 1.3 A simple analogue recording process. Although the wave changes from audio to electronic to magnetic, the signal remains analogue throughout the entire recording and playback process.

replayed, amplified, and sent to a loudspeaker so that the sound originally captured by the microphone can be heard.

Digital audio in one form or another has been available for some time: digital audio tape (DAT), compact disc (CD) or digital radio (DR, previously known as digital audio broadcast or DAB) all use digital data to store or transmit audio information. In order for listeners to hear this information, data are turned back into analogue audio within devices called digital to analogue converters (DAC).

Sampling

In order to create a digital representation of audio, the analogue audio signal is sampled. Sampling is a typical method for representing analogue information in a digital format and involves measurements taken over time.

PCM uses pulse amplitude modulation (PAM) as a means of modulating an analogue signal into a series of pulses. The analogue signal is split into small time divisions, and the amplitude within each division results in a value depending on the scale. The scale depends upon the size of byte used to store the number and common word sizes include 16 and 24 bits amongst others.

The number of time divisions per second is known as the sample rate, and this is given as a frequency. Common frequencies are

44 100 Hz and 96 000 Hz. During the analogue to digital (A/D) conversion process, each PAM pulse is converted into a number and placed within a digital word of fixed length.

The interval in use during the sampling process is most useful when it has been documented within a standard agreement, and this will include details of other parameters, such as how many bits are used to store the amplitude. A number of international standards specify sampling rates that can be used for interchange purposes. For instance, household chart-topping CDs comply with the Sony Philips Red Book CD Audio Standard. ISO 10149 is the International Standard for the manufacture of CD-ROMs and CD-Audio Discs, and is the equivalent of the Sony/Philips Red Book.

These documents detail the use of a sampling rate of 44 100 samples per second (44.1 kHz). There are two sound streams recorded on the disc, one for each speaker in a stereo recording and since a CD can store up to 74 minutes of music, the total amount of digital data that must be stored on a CD is:

44 100 samples/channel/s × 2 bytes/sample × 2 channels

× 74 min × 60 s/min = 783 216 000 bytes

Additionally, other data stored on the CD, such as table of contents and the full details, can be found in the standard. Therefore, the actual number of bits stored on the CD is a little higher than this.

The Red Book also details the size of the word into which the measurements can be stored as containing 16 binary positions. The maximum number that can be represented by two states occupying 16 binary positions can be calculated as 2^{16}, or 65 536.

When drawn on a graph, the time is represented on the horizontal axis (x), and the amplitude is represented by the vertical axis (y). For each position along the time axis, the amplitude of the analogue signal is measured and the value is placed within the word and stored.

A more straightforward illustration is shown in Figure 1.4(a), as a piece of graph paper with the lines plotted at 1 cm intervals. If a complex wave was drawn on the paper, using amplitude and time as the axes, then for each column of squares, a value could be assigned depending upon where the wave crossed the line. The values are written down and carried to a new location, where the process is reversed. The result shown in Figure 1.4(b) is a blocky representation of the original wave, as shown.

Figure 1.4 Sampling principles. (a) Pulse code modulation measures the amplitude periodically at the sample rate. (b) Whole numbers stored during the sampling process are used to reconstruct the wave.

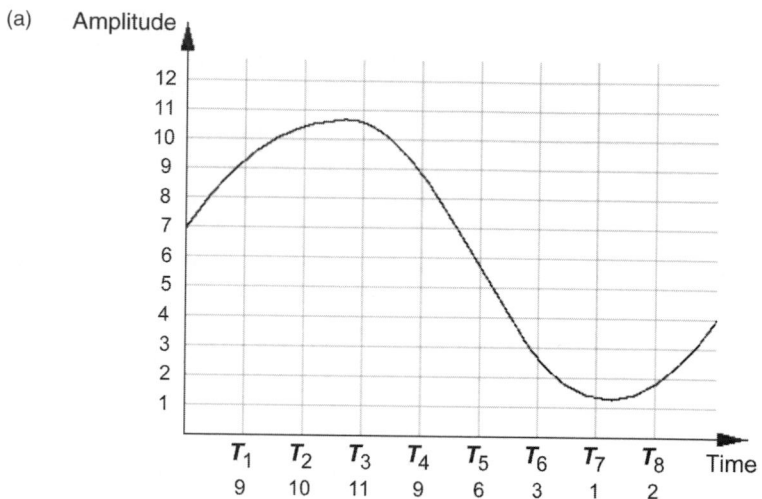

(a)

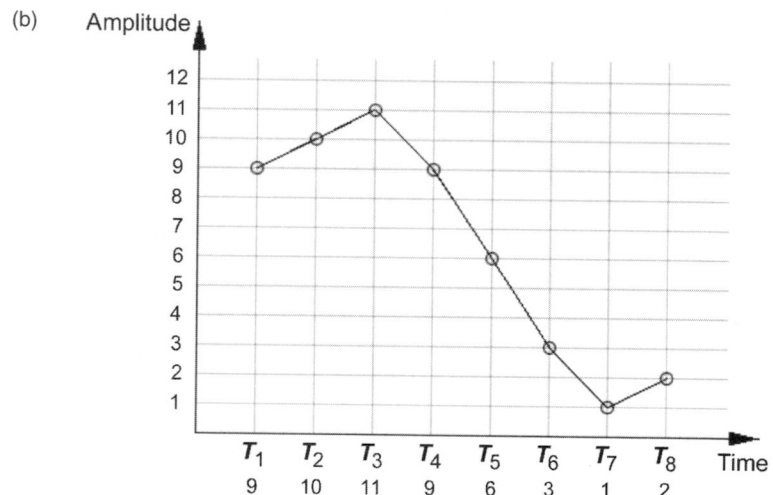

(b)

If more accuracy were required, then the squares on the original paper are made smaller, which increases the resolution. The result would be more values along the time axis, and a greater range of values for the amplitude axis. The result is still blocky but nonetheless a more accurate representation of the original wave than the previous low-resolution version.

Resolution

The combination of time and word length creates a value known as resolution. Both axes can be increased or decreased in accuracy. For instance, the word length might be increased to 24 bits, allowing the representation of a greater range of whole numbers ($2^{24} = 16\,777\,215$). The frequency with which samples are taken may also be increased (to 96 kHz for instance, which is an increasingly common professional audio format) and a combination of these would commonly be written as '24-bit 96 kHz' or 24/96. This combination is often referred to as high resolution or hi-res for short, but is not the only combination that might be known as such, since other higher resolutions have already been ratified as standards and many combinations can be imagined.

Since the number of states that can be represented in a longer word length is higher, the amplitude can be more accurately reproduced. This allows the dynamic range of a signal to be represented with greater accuracy since a higher range of whole numbers can be represented for each amplitude sample, giving more scope for slight variations (see Figure 1.5).

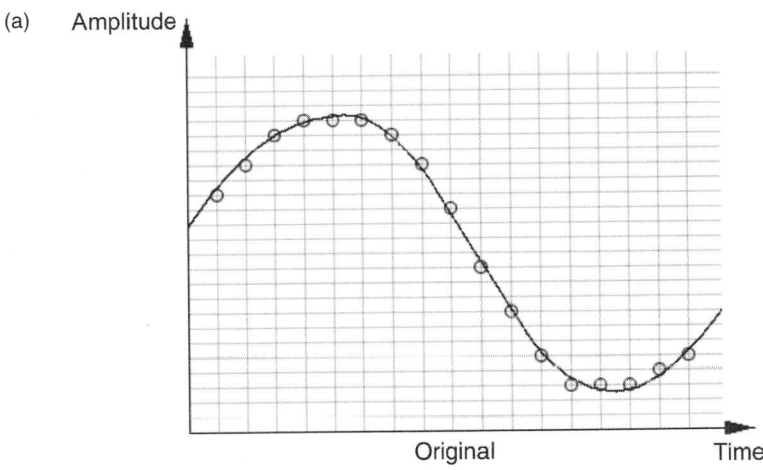

(a) Amplitude / Original / Time

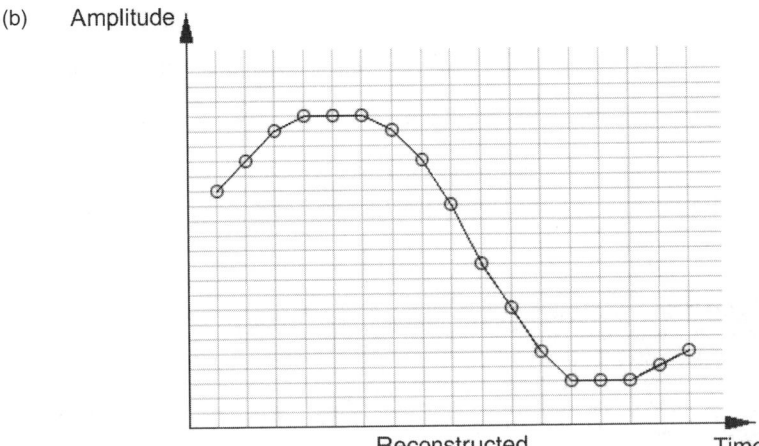

(b) Amplitude / Reconstructed / Time

Figure 1.5 Resolution. Increasing the space in which data can be stored means that a greater quantity of whole numbers can be represented. This increases the accuracy of representation of the amplitude. This is shown by smaller scaled steps in the *y*-axis. Similarly, increasing the sampling rate increases the accuracy in the *x*-axis.

Using the graph example, the accuracy of the reproduction is increased by drawing the lines on the graph closer together. The results will still be step-like compared to the original using only this technique, but at some point, the reproduction will be accurate enough for the purposes.

For example, if you were to walk around the coastline of a large desert island pushing a wheel measuring exactly 10 m in circumference, then the number of turns of the wheel would give a value for the distance travelled.

If the same walk were performed using a wheel measuring just 1 cm in circumference, then the value for the length of coastline would be completely different. This is because the wheel measuring 1 cm will measure the distance around outcrops and crevices that the 10 m wheel could not 'sample'. At some point, there is an agreement that a particular scale of measurement is sufficient for the purpose.

Since a 16-bit word length means that the range of numbers that can be stored is 65 536, this is the number of different amplitudes that can be stored. The process includes a quantization activity, which forces any fractional numbers into whole numbers, as illustrated in Figure 1.6.

As can be seen from the diagrams, a step-like or blocky representation of the original signal prevails at this point. Because the ear is more used to hearing smooth transitions in audio, the waveform is reconstructed by passing the samples

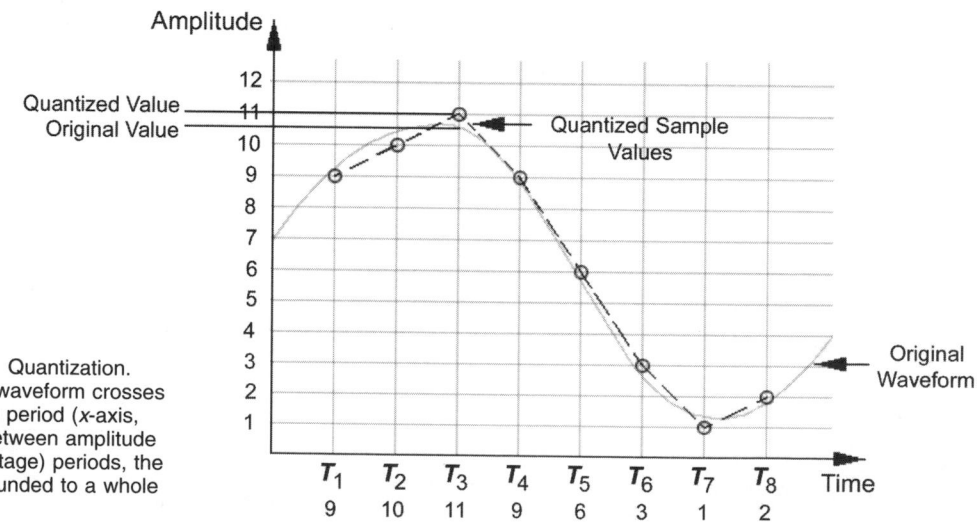

Figure 1.6 Quantization. When the waveform crosses a sampling period (x-axis, time), in between amplitude (y-axis, voltage) periods, the result is rounded to a whole number.

through a low-pass filter and this provides a kind of interpolation activity between the values in each time position (see Figure 1.7a).

A more formal process of interpolation is used to conceal errors such as missing samples. In this case, the value of the missing sample is assumed to be the average of the midpoint on either side of that sample on the time column. Instead of reproducing this value, and then immediately following this by reproducing the next value in the adjacent time column, a line is drawn between the two and a suitable adjustment to the output signal is made during the time in between the two values (see Figure 1.7b).

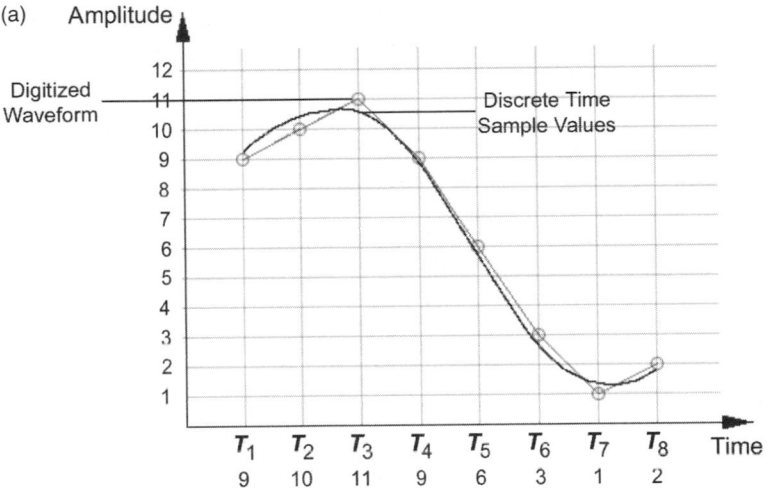

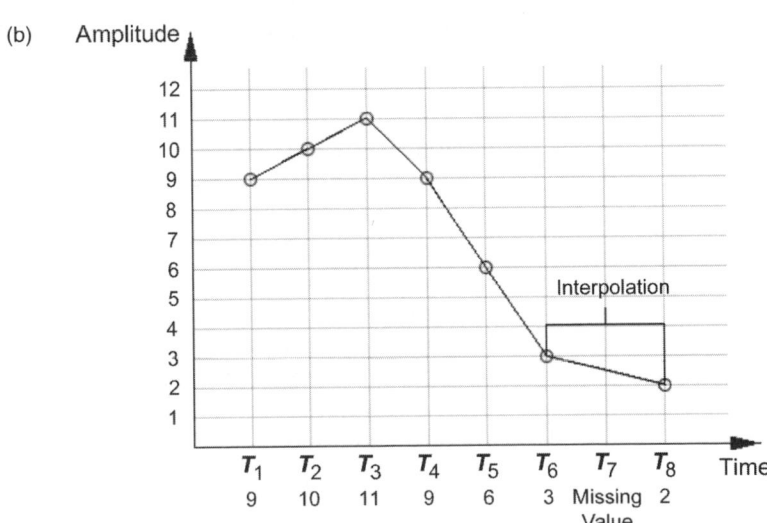

Figure 1.7 (a) Reconstruction. (b) Interpolation. Errors resulting in missing values are handled by assuming a value at that sample period, based upon the values in adjacent sample periods.

Error correction techniques used during the transmission of data over a communications channel are covered in more detail within the relevant sections.

1.2.2 MIDI and control data basics

Although analogue audio data sampled into a binary representation is the data type responsible for the bulk of data that audio production generates, perhaps the most common type of data in a wider view of computing systems is that used for control data.

The most obvious and probably the most common type of control data in use within the audio industry is MIDI data. MIDI stands for musical instrument digital interface and provides a means for the transmission of data that can be used to remotely control a MIDI enabled instrument.

It is not the purpose of this book to provide a complete understanding of MIDI and its inner workings but a description of its purpose is provided here as an overview. Readers comfortable with MIDI data and its purposes may move to the next section or if interested in a more detailed guide, are directed to *MIDI Systems and Control* by Francis Rumsey (1994).

The principle is simple to understand and is used throughout a number of industries where remote control of a device is required. In the same way that the analogue to digital (A/D) conversion must be exactly reversed during the D/A phase for it to be of any use, any analogue information can be digitized, provided the rules are agreed at both ends of the process. For any process, it is important to identify the parameters to be controlled in order to reproduce the required behaviour.

Once the parameters have been identified, a signal representing a particular parameter or behaviour is sent to a remote device where it is interpreted and the appropriate function performed. In concept, this is like creating an alphabet or language for the device. An example would be the control of a radio-controlled toy or robotic arm. In such an example, a signal is sent from the hand-held controller to a toy car, and causes the car to steer one way or another, or move forward or backwards, thus creating a control language. In practice, such toys generally use analogue signals to represent the movement of a solenoid within the toy. However, there is no reason why the process could not use digital representation.

Supposing the movement of the toy's accelerator was split into 128 small sections, as illustrated in Figure 1.8. If the solenoid were

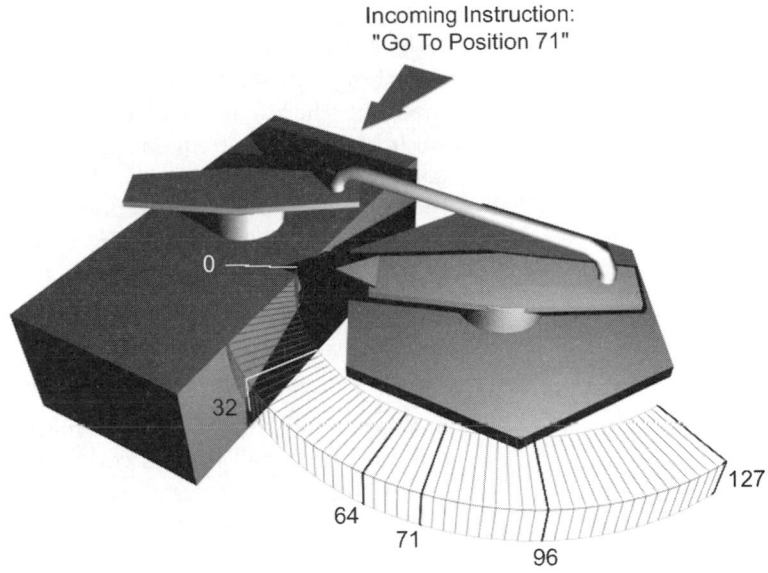

Figure 1.8 Representation of control: a digitally controlled solenoid responds to the incoming message by moving the motor to position 71. This in turn increases the voltage to a motor, making the toy car accelerate.

to move one section, then this is 1/128 of the total distance that the solenoid can move. The instruction that is sent from the hand-held controller could take a number of different forms. The signal could represent +1, meaning that the solenoid moves 1 place forward, and so the car will accelerate slightly.

On the other hand, the signal could be interpreted as an instruction to the solenoid to move to an absolute section within the range. Therefore, if the solenoid is in position 70 or any position, then the sending signal will be interpreted as 'go to section 71'.

Although it is possible to control the car using either technique, it should be noted that if the starting position cannot be determined, only the second example will work correctly every time.

Once the issue of what to measure has been taken care of, the only other problem is the matter of ensuring that this instruction is translated correctly into the analogue movement of the solenoid at the other end.

Instead of transmission over radio waves, MIDI devices respond to digital control signals sent down a cable connection. The advantage over analogue control systems is that signals controlling different parameters or devices can be identified with a digital signature or address.

Returning to the radio-controlled car metaphor for a moment, there is a need to control both the steering and the acceleration. If the signal for both control signals is sent over an analogue medium, then it is usual to separate the signals by using two slightly different frequencies. With digital data, both sets of control signal can be sent on the same frequency and instead of receiving a simple 'go to section 71' signal, the signal would more likely represent 'solenoid 1: go to section 71'.

Such a system transmits control signals one after the other, known as serial mode, although if the data are split into small enough packets, and sent out quickly enough, then the receiving device will appear to respond instantly.

In this way, a large number of unique control signals can be sent out, and provided each is correctly interpreted at the receiving end, complex control sets can be built up that are capable of controlling multiple radio-controlled cars simultaneously. This technique is used throughout different industries and various standards exist depending upon the circumstances and type of the equipment in question.

The work in writing control standards results in a uniquely identified instruction set so that instructions can be understood by various manufacturers' equipment, which are thereby made to be inter-operable.

The MIDI standard (see notes and further reading, Standard MIDI-File Format Spec. 1.1) is an impressive achievement; before MIDI, there was no universally accepted method of connecting and controlling musical instruments (primarily keyboards and drum machines) remotely. Previously the most widely used method was control voltage (CV) and gate, which used a DC current proportional to the pitch of the note being played, along with a trigger signal. The problem with this technique was that it was barely extensible, and further more complex control signals were difficult to achieve within the limited range of available signals. Additionally, different manufacturers used different voltages, meaning that commands could not be interpreted between different manufacturer devices.

MIDI was developed in the early 1980s, and used a system of byte-sized commands. Although MIDI is taken for granted today, the task of identifying the necessary commands for musical control was not an easy one, and became the subject of much study. In the end, a system of control was identified and it is testament to the work that it remains intact today, with the exception of extensions made on the standard, as demands on the technology increased and requirements progressed. The main principle of the MIDI standard is based upon a few controls.

MIDI commands

For the purposes of this book, the important controls to understand are note-on, note-off, velocity, and master volume. Several other controls are defined, and space is left within the standard for further assignments to be made by the manufacturer.

These manufacturer-specific commands are known as system exclusive instructions, or sysex for short. The only other important aspect that will be dwelt upon is the concept of channels.

Note-on and note-off are almost self-explanatory. A signal is sent out from the controlling device, or master, telling the receiving device, or slave, to sound a particular note. On a synthesizer, this will be a note of a particular pitch as if it was played from the keyboard, whereas on a drum machine, the same note will typically be assigned to a single percussion instrument.

Note-off, as the name suggests, turns the note off in the same way. Combined with note-on, the instrument will play a note, and will continue sounding that note until it receives a note-off command for the same note. There is also an 'all notes off' instruction to turn off all the sounds for the instrument simultaneously. The instrument will play a note, in response to a 'note on' command, at a volume determined by the value given as 'velocity' which is so called as it represents the velocity with which a key on a musical keyboard is depressed. As with other MIDI commands controlling aspects of the sound, velocity has a value range from 0 to 127. The range is chosen because of the ease with which it can be represented in binary form. In fact, the byte is 8 bits long, but the first bit (MSB) is assigned as a status bit used to represent the type of MIDI data. This leaves 7 remaining bits for the value, giving a possible range of 2^7 (128).

In this way, a number of notes can be individually represented, and turned on. These can also be turned off again in an order independent from the order in which they were turned on, meaning that more than one note can sound at a time. Master volume and expression allow overall control of the volume of the instrument, allowing it to be controlled proportionally to and independent of the velocity settings for each note. This means that louder and quieter passages of audio can be represented without the painful process of resetting the velocity values for each individual note.

Topology

The concept of communication channels is an important one, as it is the means of identifying different instruments on the same cable. This can be likened to controlling more than one toy car by identifying a receiving channel. MIDI cabling allows instruments to be chained together one after the other. Each MIDI device contains one, two or three MIDI connections or ports, labelled as in, out or thru (sic). The in port receives the incoming signal and passes it to the instrument where the data are acted upon accordingly. The thru port takes a complete copy of the information being received at the in port and sends it to the thru port for onward transmission to other devices in the chain. There may be more than one thru or out ports on a device. The out port takes the information generated only by the instrument itself, such as may be created when a key is pressed, and sends this onward, effectively ignoring the data from the in port.

Channels

Different instruments on a MIDI network are recognized by designating channels. This part of the system uses 4 bits of any data transmission to identify a channel. In the original MIDI specification, 16 channels were defined (2^4).

The result of a master device sending out a MIDI note-on request to channel one, using this system, is that only those instruments set to receive on channel one will respond to the request. Although another device in the chain may receive the data, the requests are ignored, because it will identify that it is not the intended recipient, as it is set to receive data on a different channel.

Illustrated in Figure 1.9, this is an important concept to understand, since it is a simple example of addressing which will

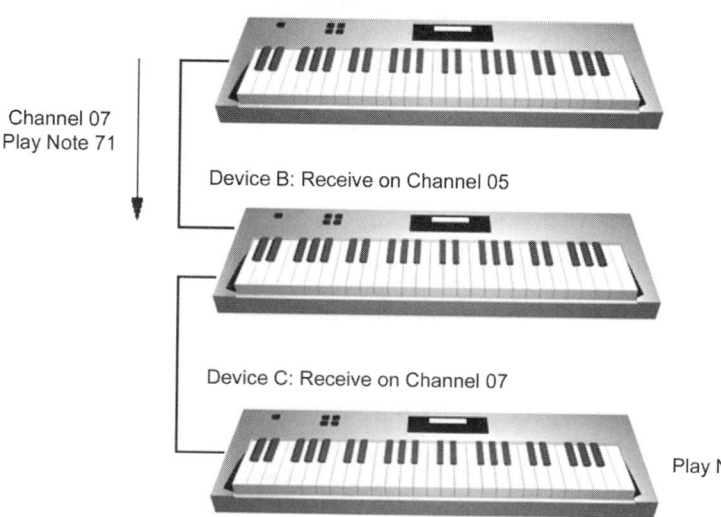

Figure 1.9 A simple MIDI network illustrating the concept of addressing, known as channels. Device A sends out a message to channel 07 to play note 71. Device B is set to receive on channel 05 and so ignores the message. Device C is set to receive on channel 07 and responds to the message by sounding the note designated as note 71.

Device A: Master

Channel 07
Play Note 71

Device B: Receive on Channel 05

Device C: Receive on Channel 07

Play Note

be covered in greater detail throughout this book. Packet networks, used as a common form of computer communication, identify the intended recipient on a more complex network in a similar way, by attaching address information to the front of a data packet. The key point is that all the control data travel down the same cable and are identified only by the contents. MIDI refers to each address as a channel.

1.2.3 Timing and synchronization information

The third type of data worthy of examination with a view to understanding the issues surrounding transfer is that of timing and synchronization signals. These are two of the most important aspects of connecting audio equipment together, particularly in multimedia and visual applications where the audio signal needs to be closely synchronized with the visual element of the production in order to attain the best experience. The issue of synchronization is also important in multi-track recording environments, especially when multiple sessions or multiple studios might be involved in the audio production process.

Timing

Each time audio material is changed, added, or removed, it is desirable to ensure an understanding of time compared to the rest of the session. If correct timing is not maintained, a particular audio snippet may appear to be out of time. Several types of timing and synchronization signals exist in order to allow the transfer of this understanding of time.

Timing is intrinsic to digital audio since data are grouped into words that are replayed over a specific time duration. Therefore, it is important that all equipment within a studio that is designed to playback digital audio should have a common interpretation of time.

Control data can also be related to time in the same way, and it is natural that MIDI has its own clock signal. This clock signal is intrinsically related to the tempo of a musical piece, and so cannot be used as a master clock for synchronization to more complex audio production, which might not involve music.

In the case of A/V facilities, a synchronization signal such as devised by the Society for Motion Picture and Television Engineers (SMPTE) may be used, or an internal master clock may be generated from one of the available commercial devices designed for the job. A master clock will often only produce a pulse, which can be translated into other clock and reference signals depending upon the requirements.

1.3 Why digital audio?

Digital data offer advantages over analogue within the audio production process because the normal interference and degradation, that is so much of a problem in analogue forms, do not affect them in the same way. The normal audio production process involves capturing a performance or audio of some description and performing some processing upon it such as mixdown. Normally the final result will then be stored or broadcast.

The processing may include microphones, cables, mixing consoles, effects processors (such as equalization, delay, compression and so on), amplification, and a storage medium such as tape or disk. Each device or process may introduce some form of noise to the signal, no matter how small, through interference caused by radio frequencies (RF) or by other electrical problems, such as a poor earth, incomplete connections or simply the presence of other electrical devices.

The specific problem with noise in the analogue domain compared to the digital domain is that noise in the analogue domain cannot be easily separated from the original signal. Noise added to one valid voltage results in another valid voltage with the result that it is almost impossible to ascertain what proportion of noise has been added.

With binary representation, there are only two possible states that the signal can be in, and since these are predetermined, any voltage that falls outside of the expected limitations can be recognized as an error and handled through corrective measures. Fluctuations may manifest themselves as unexpected peaks or troughs in the waveform, which can be handled within the system design, usually by an electrical circuit designed to detect and respond to electrical states and fluctuations. Other discrepancies may appear on the timing boundary between bits, thereby effecting the y-axis as well, and these may be handled by timing and synchronization methods and so not result in loss. In this way, the digital signal is far less prone to the types of distortion and noise that normally spoil an analogue signal.

The amount of noise that can be tolerated is determined by the type of binary representation in use, for instance high and low voltage signals used for representing 1s and 0s will have expected voltages for either state. In this case, adding noise will result in an invalid voltage, which can be identified and dealt with separately (see Figure 1.10). There are several mechanisms in use to make digital signals more resilient to errors of this kind, which are discussed later in this chapter.

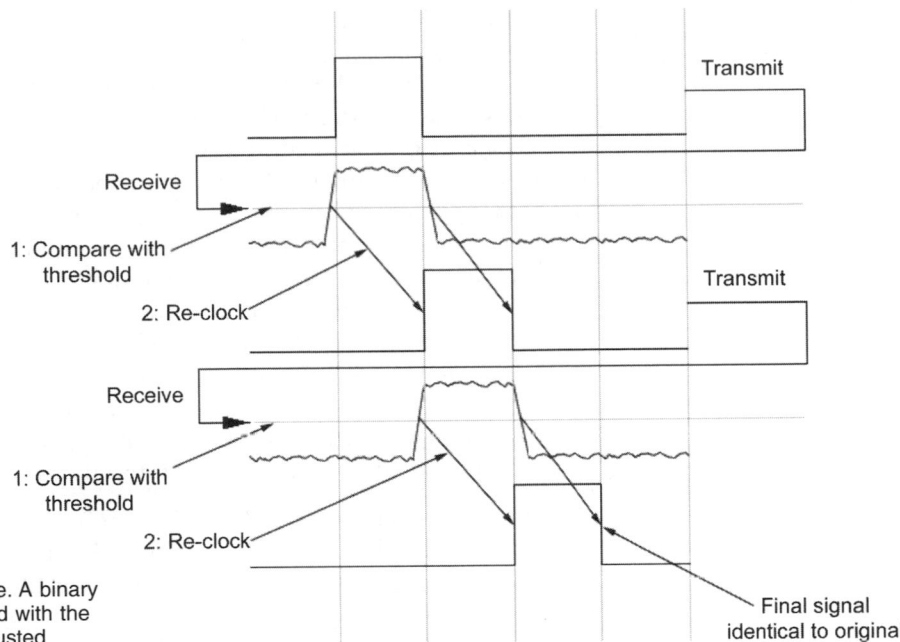

Transmit

Receive

1: Compare with
threshold

2: Re-clock

Transmit

Receive

1: Compare with
threshold

2: Re-clock

Final signal
identical to original

Figure 1.10 Noise. A binary signal is compared with the threshold and adjusted accordingly. The interested reader may also find references on jitter useful. Source: *The Art of Digital Audio* by John Watkinson.

Digital audio environments also offer excellent convenience of manipulation since editing need not take place linearly, as might be expected when manually splicing a tape to move a particular snippet of audio. Instead, the edit can take place within computer software, and an accurate edit of a single track in a multi-track recording need only take a fraction of the time. This advantage has been exploited to excellent effect by offering audio and visual non-linear editing over multiple tracks and sessions within the digital realm, increasing productivity compared to analogue manipulation.

It should be considered that while the signal is moved or manipulated in analogue form, and each time the signal is converted from analogue to digital (or back again), there is the possibility that some additional noise may pervade the signal. Therefore, once the decision has been made to use a digital production process, it is preferable to keep the information within the digital domain as far as possible.

Digital formats are favoured within the broadcast environment such as radio and television, where facilities are switching to broadcast the digital information direct to the consumer, whose equipment performs the decoding back to analogue.

There are a number of reasons that digital information is preferred over analogue in this instance. The resilience that digital decoding techniques offer and the ability to multiplex channels together are two examples. Interference generated by reflections from nearby structures (such as buildings or natural geography) or weather conditions can be removed during the decoding process, since the decoding mechanism can be programmed to ignore fluctuations, variations and additional noise that force the signal outside of its normal pre-programmed parameters. Although this is true to a certain extent, extreme conditions can still degrade the signal but overall, the transmission of digital signal offers better resilience than the analogue counterpart. The receiving device can interpret the digital information, and not the distortion common in analogue signals caused by reflections or weather conditions.

Another advantage in using digital signals is the ability to multiplex or combine signals together. This means that several completely separate signals can be sent over the same communication channel (such as cable or a particular radio frequency). This can occur by some means of addressing or recognition of the components of each stream within the broadcast. This is a complex subject and is covered later, but means that independent streams can be broadcast over the same communication channel, sharing the data rate between them. Early uses for this are in the broadcast of events such as concerts, where it is possible to broadcast multiple audio mixes. For instance it might be possible to listen to the monitor feed experienced by a member of the band, or to provide multiple camera angles, such as is in use within the coverage of sporting events.

From a quality point of view, such technologies mean that the consumer device – such as the CD player or digitally enabled television or radio – can perform the translation from digital to analogue audio stream, extending the digital domain all the way into the home.

1.4 How does digital technology work?

To explain how digital technology works, it is appropriate to look at the example of a popular general-purpose computing device such as a personal computer (PC). The PC is an example of a microcomputer, so called because most other types of computers are bigger, if not physically, at least in terms of power and cost.

For the purposes of this book, the PC is used as a general term to include PC proper, which is a derivative of the open IBM integrated systems architecture (ISA) hardware standard, and the Intel x086 processor series. This combination is popular in commerce as well as at home, although the term is extended to include Macintosh architecture from Apple, popularized within the home and in businesses such as audio production, publishing, and multimedia. Finally, PC can also be used to describe various digital audio workstations (DAWs) designed specifically for the manipulation and storage of audio within the digital domain.

This section explains terms and concepts associated with computing that are used throughout the book.

1.4.1 Personal computers

Computers are made up of three main components: the central processing unit (CPU); memory; and some sort of storage device, commonly a hard drive (or disk), although this was not always the case. Other important components include some way of viewing and inputting data and these are grouped together as input/output devices (I/O). In the case of the PC, input is binary information translated from a keyboard, mouse or audio signal and the output is the result of processing according to some computer program. The relationships between components of a PC are shown in Figure 1.11

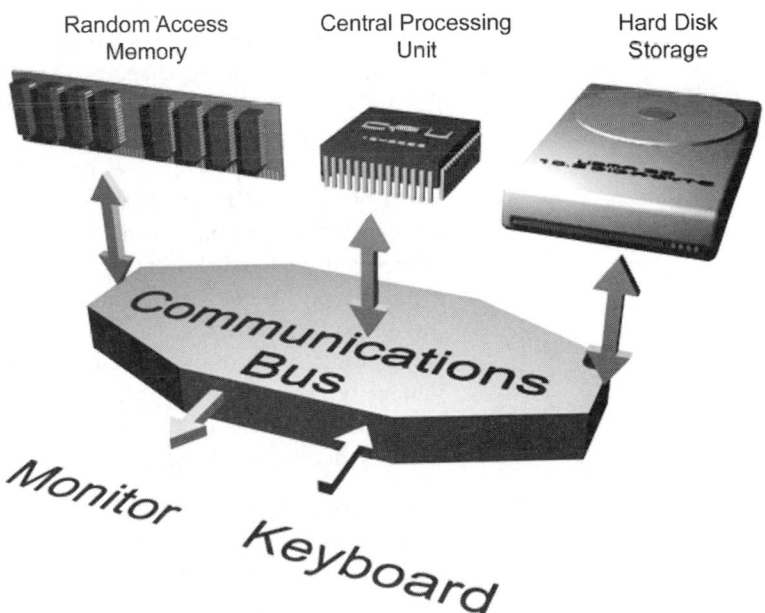

Figure 1.11 A simplified representation of communications between the components of a PC, illustrating a bus through which all communications take place.

1.4.2 Central processing unit

The CPU is the 'brain' of any microcomputer and in many cases performs the actual transformation between input and output. In PCs, the CPU is usually a single silicon chip containing millions or billions of transistors designed to perform logical functions, and as the name suggests, does the calculation and processing.

Trends vary depending upon technological progress and it is possible to move specific calculations for tasks such as audio onto chips designed specifically for the purpose, such as digital signal processors (DSP). Essentially, this is like delegating the task to equipment more suitable for the job. As technology and commercial forces dictate, this task might be brought back onto the CPU. Intel's MMX instruction extension to the Pentium CPU series was an example of offering additional media services on the main processor, and quickly became standard.

CPUs can also route data from one place to another, and this functionality can be likened to parts of a mixing console, where the output from one channel may be routed to the input of a sub-mix or bus.

1.4.3 Storage

Once information is rendered into digital form, it needs to be stored, or all the processing and programming is for nothing, since it will be lost when the machine is turned off. Storage devices store information so that it can be retrieved at any time. The most common of these is the disk. These are called either floppy, or hard disks. The term floppy originated in the days before magnetic media came in any kind of plastic housing, as with today's floppy disks, or ZIP drives and refers to the flaccidity of the thin circular media. Hard drives (HD) are often located within the main unit of the device or within a disk array, whereas floppy drives tend to be portable in nature. Hard drives are used as the main permanent storage media within PCs, and the amount of data that can be stored upon them is increasing as technology progresses. Floppy disks on the other hand are generally used to carry information from one PC to another or as an effective means of temporary storage for a small amount of data, typically 1.44 Mbits.

Other types of storage mechanism use other ways of representing 1s and 0s, such as in CDs, which use the presence of a microscopic dent or the absence of a dent (known as a land) to represent the two states.

Commonly, disks and tapes use magnetic (or optical) means in order to store binary data, in much the same way that cassette tapes are polarized to store analogue information. These types of storage devices are called persistent, since they retain their information, even when the machine is turned off.

1.4.4 Memory

Memory is another form of data storage, using silicon to store information, allowing faster access than is associated with disk or other mass storage, and memory can take several forms. RAM (random access memory) is probably the most commonly thought of memory type and this is temporary in nature, since it loses its information as soon as the device is turned off.

For memory to work, a matrix is built up to deliver electricity to silicon capacitors that either store a charge or do not, depending upon their binary state (1 or 0). ROM (read only memory) works in the same basic way but is persistent and so keeps its settings after the device has been powered off. Electronically programmable read only memory (EPROM) cannot generally be altered by any user functions, and needs a special 'EPROM blower' before it can be loaded with new information, and often contains program code but does no processing. This type of memory is commonly used to store the code for video and arcade games or for the basic input/output system (BIOS) functions within a PC. BIOS functions are some of the very lowest form of code for a microcomputer, and describe the basic functions upon which the computer will build. For instance, BIOS identifies where the keyboard input will come from, and what shape will be displayed on the screen when a letter is pressed on the keyboard. BIOS also points to the area where the next layer of information or operating system can be found (usually the first track on a hard disk).

1.4.5 Communications bus

A bus is a communications channel that allows the movement of data from one area to another. For instance, the operating system may read some commands typed in on a keyboard that instructs the machine to retrieve some information stored on disk, and send it to some output device, such as an audio card. In simplified form, the request is sent to the disk, which returns the specified file down the bus, as shown in Figure 1.12. In non-bus-mastered systems the information passes from the disk and onto the bus, and then to the processor. The processor then routes the information back down the bus to the audio card.

Figure 1.12 Simplistic data storage and retrieval. Stage 1: a user enters some data or commands, which the CPU interprets according to the rules loaded from the operating system. The result is a request made upon the disk for a file to be retrieved. Stage 2: the disk receives the request from the CPU and responds by returning the file, whereupon it is stored in memory.

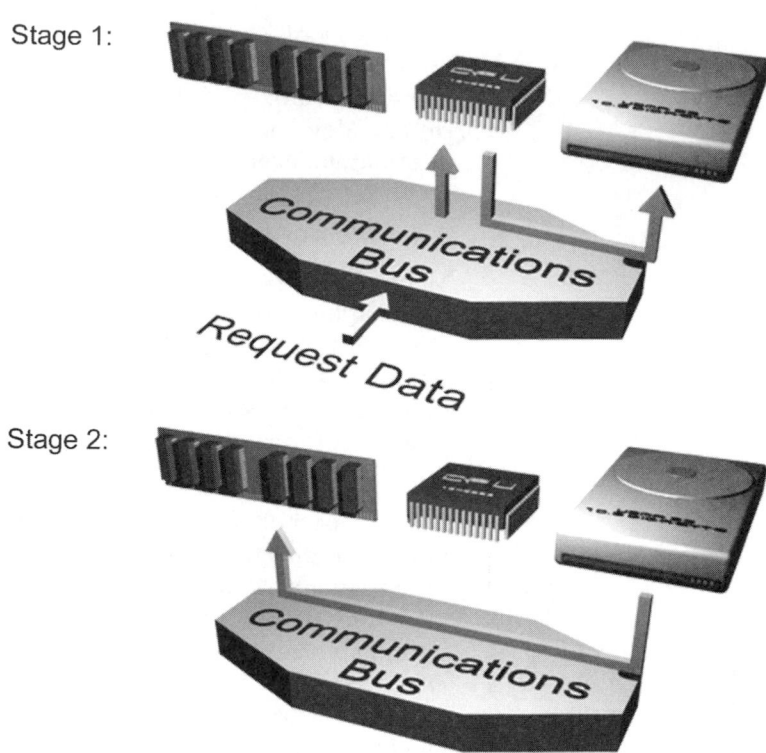

Stage 1:

Communications Bus

Request Data

Stage 2:

Communications Bus

In the same way, an input device such as a keyboard can send a signal, which can be interpreted by the CPU. The CPU can then process the information it receives, and route the result to another device such as a screen. In this way, words appear on the screen when typing.

Peripheral component interface (PCI) is one of the most common bus types, and is interesting since it was one of the first to offer bus mastering. Bus mastering is a technique whereby components attached to the bus need not use the CPU in order to talk to other peripherals on the bus (provided the other peripherals support this function) thus alleviating the CPU of routing tasks.

1.4.6 Binary encoding

Binary encoding techniques describe exactly how 1 and 0 are represented. As mentioned, the simplest form of electrical encoding would be for a voltage to be turned on or off. This might be true within computing devices, but when data are moved around, some other points need to be considered so that data can be moved around and understood by all parties successfully, a little like the signs used in the highway code to ensure order on the roads.

In electrical binary devices such as computers, there are generally two different voltages, referred to as the high signal and the low signal because one voltage will be higher than the other and will appear higher or lower when drawn on a graph. In practice, these signals may be two different voltages on a copper cable or two different light intensities on an optical link.

When two digital devices are attached together using some form of communication system, there will quite often be a synchronizing clock signal accompanying the 1s and 0s, which ensures the two stations are working at the same speed.

Problems occur with such a simple mechanism, however, when a whole series of the same state are sent consecutively. For instance, if the high state represents a 1, and there is a requirement to send one thousand high states one after the other, then there is a good chance that the continuing, unchanging voltage will be difficult to accurately assess across time at the receiving station, without any reference to the sending station's understanding of time. The receiver is likely to lose count of how many bits have been sent.

This problem is apparent in the simple non-return to zero scheme (NRZ; see Figure 1.13a), but can be overcome by representing a logic bit by a transition between voltages, rather than the high or low state itself. In this way, a long string of the same binary state will be represented by changing or transient voltage. The Manchester encoding technique is similar to NRZ, but uses a state change in the middle of the bit period so a 1-bit is transmitted with a high voltage in the first period, and a low voltage in the second, and vice verse for the 0-bit (Figure 1.13b).

Another scheme, Differential Manchester, represents a 1-bit by making the first half of the signal equal to the last half of the previous bit's signal and a 0-bit is indicated by making the first half of the signal opposite to the last half of the previous bit's signal. That is, a zero bit is indicated by a transition at the beginning of the bit. Figure 1.13c illustrates non-return to zero inverted (NRZI), which is the opposite of this, whereby a 1 is represented as a transition, and a zero is represented as a non-transition.

1.4.7 Networks, stations and nodes

A network consists of two or more computing devices that are linked in order to share resources (such as printers and CD-ROMs), exchange files, or allow electronic communications. The

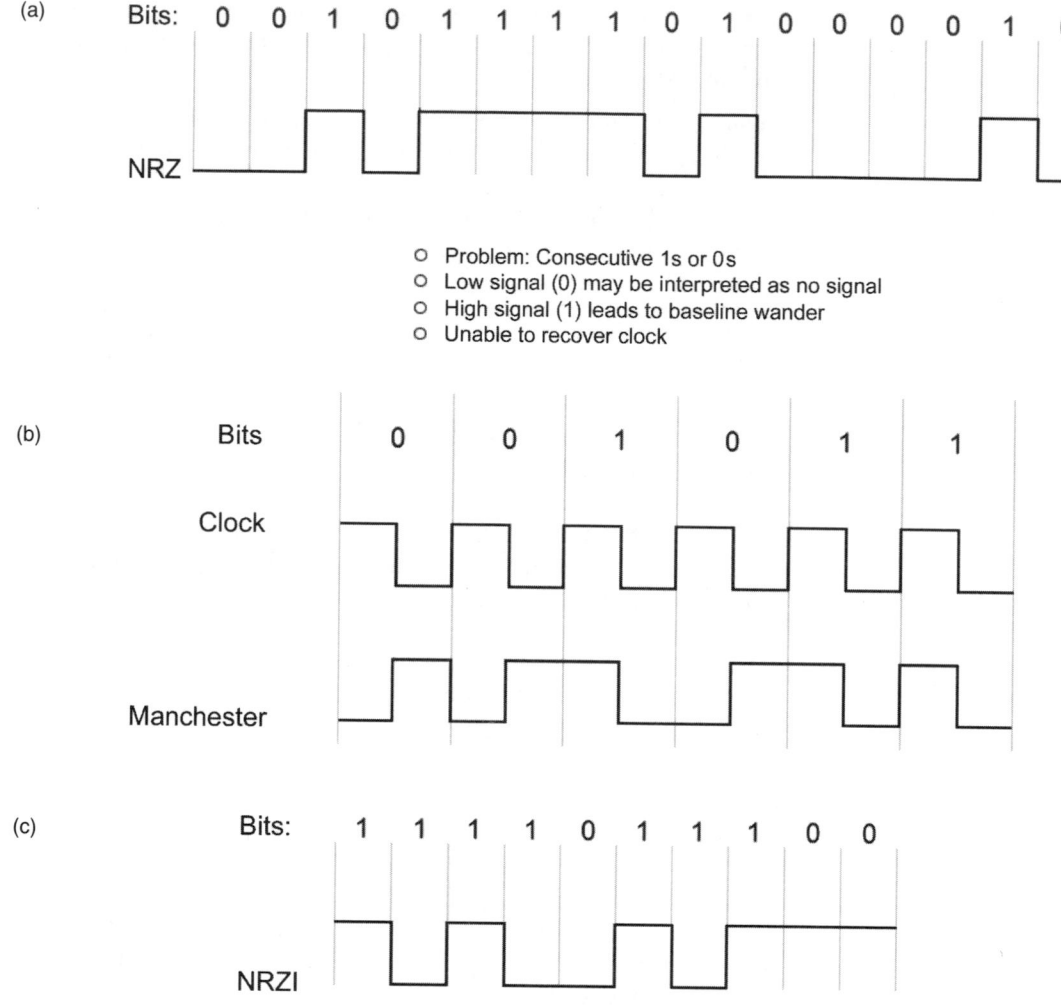

Figure 1.13 Encoding binary states onto a media. (a) Non-return to zero. A logic 1 is sent as a high signal, and a logic 0 is sent as a low signal resulting in problems of synchronization when a signal contains long strings of 0s. (b) Manchester (biphase level) encoding is a synchronous clock encoding technique in which binary data are transformed so that a downward transition represents a 0 and an upward transition represents a 1. (c) Non-return to zero inverted (NRZI) uses the presence or absence of a transition to represent binary states.

computers on a network may be linked through cables, telephone lines, radio waves, satellites, or infrared light beams. There are many ways of categorizing networks, and one of these is by the geographical area covered by the network. In information systems (IS), the basic categories in this case include:

local area network (LAN)
metropolitan area network (MAN)
wide area network (WAN)

A LAN is generally installed within a single building or campus, whereas a WAN extends the concept to cover connections between towns, cities, and countries. A MAN is a newer term that can have two possible variations on meaning. The first is a network installed internally to an organization, where the parts of the organization remain within a municipal area. The second interpretation is a high capacity general network, offered as a service within a town or city, probably connecting to the Internet. Each of these types of network needs to consider different design aspects.

In network terms, a station is considered to be a computer, whereas a node may also be a network element, such as a router or bridge. Routers, bridges, and other network elements are covered in later chapters.

More recently, other network abbreviations have become popular, such as SAN, for storage area network, and CAN, for controller area network (often associated with the automobile industry). Those that are pertinent to the transfer of audio are covered in more detail in later chapters.

1.5 Why do we need to move digital audio data around?

In order for the full benefit of digital audio to be realized, the digital domain can be extended into the consumers' home. The signal may be broadcast, or the digital data stored on a medium and purchased in retail outlets – such as purchasing a CD. No matter how the data are moved, the receiving device must correctly interpret the digital information, otherwise the resulting 'audio' may not be at all desirable. In both the above cases, the purpose of transferring digital data is to distribute the product to the consumer.

At a more fundamental level, the requirement to move digital data has been around since computers were first invented. There is little point in designing and building a computer that cannot move data from the CPU to somewhere that the results can be used by the user, since if this were the case, the results would remain inside the CPU and could not be accessed.

Once the problem of moving the data around inside a single computer has been solved, it is an extension of this to move the data over greater geographical areas between different computers and to use and store the data in devices designed for the specific purposes. Moving data between devices opens up whole new levels of flexibility in terms of what can be done with the data, who can use it and so on. Ultimately, this paradigm extends to moving digital audio data directly between two media studios or to the consumer, in the form of files or streams. The uses to which this mobility can be put are still being explored as the capabilities of the technology increase.

Rather than cover the full spectrum of techniques employed to move digital data around, from broadcast through to retail distribution, this book focuses on discussion of the various techniques that can be employed to transfer digital data from one place to another during the audio production chain.

It is important to understand that the type of binary, digital data being transferred can be separated from the mechanism that is employed to move it. However, it is also true to say that the type of data being transferred may have a bearing on what mechanisms are suitable for successful transfer.

For instance, audio data takes several forms. It may be in the form of a file or sample stored on a computer or sampler, or it may be broadcast in traditional terms using radio frequencies in the form of a stream arriving constantly over time. Both of these examples have certain criteria that need to be fulfilled before the delivery can be considered acceptable.

Any number of mechanisms may be employed to move the data, but will often do so regardless of whether it is meeting the quality of service (QoS) required by that type of data. If the acceptable quality is not met, then the result may be described as an unacceptable quality of service. Nonetheless, the mechanism in question will still function without caring whether or not it is meeting the desired QoS, and the result may well be unacceptable loss of information over a specified time period.

For instance, during the recording process, the professional studio environment requires an uninterrupted stream of data from the start of transmission to the end. If this is not achieved, the playback will contain unexpected spaces, skips, synchronization, and timing problems. This renders the quality of the recording unacceptable for professional use or as a consumer product.

On the other hand, deliveries of samples from a database will usually only require a simple file transfer mechanism to be in place. This would not need data to be delivered in a continuous stream, since the audio is not being heard at this point, but is merely being moved from one place to another for the purposes of manipulation. Instead the file may be grouped into packets and delivered in a way that is less restrictive in time and even in the wrong order. In this case, the user will be unconcerned with the inner workings of the delivery mechanism, because the file turns up shortly after it is requested. Provided the file is pieced back together in the correct order and the delivery process completes without error then the audio information remains unaffected.

Quite often in these cases, a requested file containing for instance a three minute recording may well turn up in much less than three minutes. It would be an easy mistake to assume that the mechanism achieving this result would also be suitable for the demands of data streaming that the recording process requires.

Streaming in this context means that a fixed amount of data needs to be delivered constantly over time. The amount of data being streamed might actually be quite small, but the delivery must remain constant. On a busy network, it might not be possible to guarantee even small data rates to any particular service and device. Therefore, even though the mechanism employed to move the data may be capable of delivering a three minute audio file in much less than three minutes, it cannot sustain a given bit rate over a period of time, since a short burst of traffic from another device on the network will result in the stream being temporarily interrupted, resulting in problems with the audio quality.

Audio is just one type of data that can have different delivery requirements and there are several different types of data.

Historically, streaming of data was not one of the considerations for scientists working on the first computer networks and so it is possible to see how certain (network) industry-standard mechanisms became popular without any regard to the requirements of other industries, including the audio industry. As a result, many network technologies remain unsuitable for delivery of this type of service, although much work is being done to achieve the best results on the available technologies, and in searching out additional methods that can meet these requirements.

1.6 How are digital data moved around?

There are a great many ways to move digital data around. Some of these have been developed by the audio industry, and are in common use, while the majority have been developed in other areas, generally to do with the computers and networking.

The problems involved in building a mechanism to move data from one computer to another have been understood for some time, although when a number of different types of devices were attached to the same communications channel, the problem took slightly longer to solve.

It was not until the early 1970s, with the publication of the US Department of Defence 4-Layer Interconnection Model and subsequently the International Standards Organization Open Systems Interconnection 7-Layer Reference Model, that these steps were described in enough detail to include communication with other devices. Until that time, different manufacturers investigated and produced a wide range of different communication techniques for the purpose of communication between their own devices.

With computer networks now installed in offices and workplaces of all sizes and types, the complex mechanisms and history behind developments such as the Internet and networking in general are perhaps taken for granted. The history is important however, since inspection reveals the reasons behind the key decisions in the development of the technology, and so this is covered in more detail later.

Many different communication standards have evolved for different industries and, as mentioned earlier, MIDI is an example of one in use within the audio industry. Standards for other industries include the popular IEEE 488 standard agreement (sometimes known as General Purpose Interface Bus or GPIB, and also known as HPIB after Hewlett Packard, the company that invented it) for the control of test and general-purpose equipment, as well as Ethernet in the networking industry, which is covered in Chapter 3.

1.6.1 Serial and parallel communication

As with networks, communication techniques can be categorized in many different ways, depending upon the viewpoint one takes. Perhaps the very first decision that needs to be taken in designing a new communication standard is whether the data will be transmitted in serial form or in parallel. This has a direct bearing on how many wires are required in order to transfer the data. Parallel communication occurs when the binary digits of data are in groups, each bit in the group being assigned to a separate wire. Serial communication, on the other hand, uses just a single pair of wires in the same cable, and sends binary bits one after another, as shown in Figure 1.14. This can perhaps be best illustrated using the example of Morse code.

Figure 1.14 Serial communications. Bits are sent consecutively down a single wire pair. Annotation assumes that the leftmost bit is sent first with progression towards the right-hand side of the page.

Storage and Transmission
(Digital)

00110110010110011110111101011

Time ➤

Morse code is a representation of the alphabet made up of a series of dots and dashes. A dot is a short signal, and a dash is longer one. Since these are the only two states that are used to represent the whole alphabet, this can be considered a bi-state representation, or binary. Morse code was used for the first wire-bound transmission on 24 May 1844 rendering the phrase 'What hath God wrought' into Morse code.

For serial transmission, each letter of the alphabet is sent one after the other, and furthermore, each dot and dash is sent one after the other. If the same information were to be placed on a parallel transmission, it is feasible that each letter could travel down a separate wire, to arrive at the same time. Since parallel communication standards tend to use a fixed number of wires, then the number of wire-pairs in use represents the number of letters than can be transmitted at once. Since each letter of the alphabet is made up of a series of dots and dashes, another way to send out the same transmission would be for each binary state in one letter to be sent over a different wire. This means that each whole letter is effectively sent in the time that it takes to send one dot or dash. In binary terminology, this amount of time can be considered as one bit-period, or the time it takes to represent one state. Serial and parallel communication is shown in Figure 1.15, using Morse code as the example.

Figure 1.15 Serial and parallel transmission illustrated using representative Morse code. Note that each line of Morse code uses a separate wire pair (or communications channel). The definition of networks encompasses, by far, more serial communications than parallel because of the decreased complexity when transmitting over relatively long distances. Parallel communications are reserved for backplane environments. The principle for parallel transmission is the same as for multiple synchronized serial transmissions.

Although there are plenty of standards developed for parallel communications, these have been developed depending upon circumstances and IEEE 488 is perhaps the most prominent example.

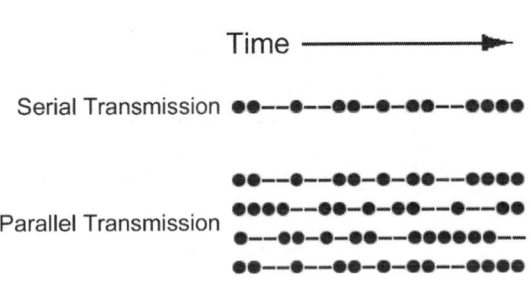

31

Most network technologies use serial communications simply because of the reduced cost associated with only using a single pair of cables. Occasionally, serial communications will still use two or more pairs of cables for different functions, such as the clock signal. Several serial communications standards have developed, depending upon circumstances, and MIDI is an example in common use throughout the audio industry.

1.6.2 Synchronous, asynchronous, and isochronous

Another way of categorizing data transfer is by the timeliness of the transfer. Technically, the question might be asked, 'Is the transfer synchronous, asynchronous or isochronous?'

A stream of bits being received needs to have a point of reference, so that each bit can be recognized, otherwise communicating systems may not understand where each bit starts and where each bit stops.

Synchronous communication occurs between two devices when a clock signal is sent along with the data. This may be sent along separate wires or modulated with the data (as per the Manchester encoding mechanism). In this way, the receiving device can interpret and lock to the clock signal directly, using it as a reference. In asynchronous communication, the clock signals of the sending and receiving devices are not locked directly, but have the same frequency. In these cases, the start of a stream is indicated by a start-bit, and ends with a stop-bit.

It is difficult for the two devices to remain exactly in synchronization without some sort of adjustment, just as digital watches may drift very slightly over a long period of time. In order to compensate for this, the clock of the receiving device is altered at the ending boundary of the start-bit, and then again at the start of the next data byte. The clock remains synchronized long enough over the period of one byte to receive the data in between (Rumsey and Watkinson, 1995).

In isochronous behaviour, the network will have a master clock signal to which all devices are synchronized. Some interesting variations are explored in later chapters, to do with isochronous behaviour, where the clock signal becomes part of the data itself, and is reassembled by the receiving device so that the audio stream can be reassembled relative to time. This kind of mechanism can only really be effective with constant bit rates (CBR) such as digital audio streams.

1.6.3 Error handling

Although digital media is often mistaken as a storage solution that does not degrade over time and one in which transmission is not affected by the same interference as analogue systems, this does not mean that digital environments are error free. The subject of error correction is as much of a science as any other subject within the computer world and is taken seriously by manufacturers of computing equipment keen to offer high availability and low error rates to customers willing to invest in the protection of their data.

Parity and checksum are two techniques used within computing environments to provide fault tolerance and error recovery. The concepts are simple to understand.

Parity

Parity or more correctly, parity bit, is an implementation whereby one of the bi-states is counted. The concept is usually as simple as adding up all the ones in a data word or byte. If the number of ones is an even number, then the parity bit is also set as a one. If the number of ones is an odd number then the parity bit is set at zero. Using this knowledge, if one of the bits is lost, then a comparison can be made, by counting up the number of ones, cross checking against the parity bit, and performing error correction, depending upon the result.

Parity bit alone cannot determine which bit is missing or corrupt, but the technique has uses allowing corrections to be made in two common examples that are in everyday use. The first is on the Internet, where bytes contain seven bits of data and the eighth bit is reserved for parity. If the parity check does concur, there is a problem with the delivery of some information and the receiving device can discard the data and request it again from the device that sent it.

Redundant array of inexpensive disk

The second example of use of parity is in disk arrays implementing the higher level RAID (redundant array of inexpensive disks) functions. A RAID array is a set of disks that are grouped together to provide a larger storage area or volume. Several levels of RAID functionality exist, the first being RAID-1, which is a straightforward copy of one disk to another (known as a mirror), which occurs on the fly, or while the system is in normal operating mode. If one disk should suddenly fail, then data will continue to be read from the other.

Levels of protection available from RAID are described up to level RAID-5 and above. RAID-5 is an implementation whereby any number of disks can be placed side by side in an array to make one large volume. The last disk in the array is the parity disk, which is considered redundant, since this disk space cannot be counted as part of the overall space available in the disk array. When data are written to the drive, it is striped across the array, so one byte is written to each disk in turn. The last, redundant disk contains parity information calculated as a function of all the bytes in the same position on all the other disks. Disk arrays that use this implementation can sustain the complete failure of any single disk in the array without losing any data or functionality. When a disk fails, it can be replaced with a new one while the system is running, without any loss of service. This is achieved by calculating the missing data by taking the data from the other disks, and comparing the new parity result against the actual parity to replace the missing information.

Checksum

Checksumming is commonly used to fingerprint, or take a snapshot of information for comparison. A typical use of checksumming is in virus protection and a simple example of how this is implemented would involve treating the whole disk as a single number, performing an algorithm on it, and then storing the result. If so much as a single bit has changed from a one to a zero, then the result of the checksum will be different when the algorithm result is compared to the original. Checksum can be performed on a single file or packet, or at predetermined intervals within a stream and several techniques are in use.

Error checking and fault-tolerance are broad technical subjects and can fill a whole book by themselves (Pradhan, 1996) and the two techniques that have been covered here are provided in simplistic form.

1.6.4 Multiplexing

Multiplexing allows more than one stream of digital information to be passed down the same communications channel. Digital streams of data particularly lend themselves to multiplexing techniques since each separate stream can contain its own unique identification, thereby separating it from the remaining streams.

Two multiplexing techniques are in common use: time division multiplexing (TDM) and frequency division multiplexing (FDM). TDM splits the available bandwidth into time slices. Each slice takes up a predetermined time period, and data from one source fills a pre-allocated division as shown in Figure 1.16. In a straightforward configuration, the adjoining time division will normally contain data from another source or stream, and will continue doing this until all the time divisions are allocated. This simple description becomes more complicated if there are empty time divisions, containing no data. For instance, asynchronous packet-like data generated by file and application requests on computer networks tend to be bursty in nature, meaning that there will be unpredictable periods of both intense activity and relative calm. During the quiet periods, there may be no data to fill some time slots, and so the slots may be reconfigured on the fly to take advantage of this spare capacity by reallocating the slot to wherever it is required. During bursts of asynchronous data, time slices will be allocated depending upon a predetermined priority, which is assigned according to how important the data are. During bursty periods, less time-critical data may be held up, so that it does not interfere with other streams.

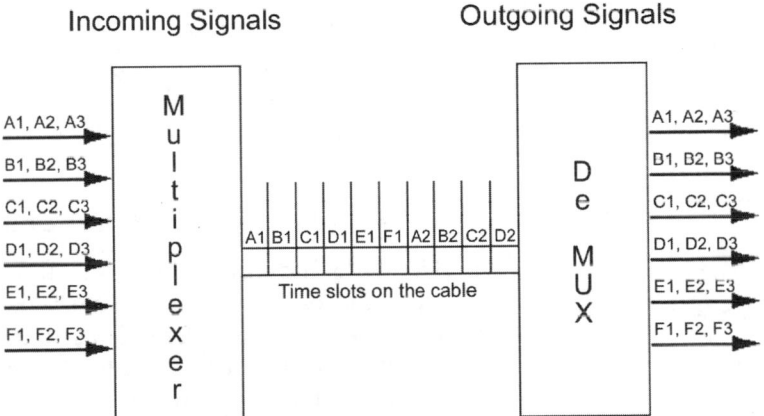

Figure 1.16 Time division multiplexing. Typically, multiple sources are given an equal slice of time to transmit data. Statistical techniques allow advanced efficiency of data rate allocation.

This allows guaranteed bandwidth for those applications and streams that are more time critical and therefore have a higher priority, such as real-time audio streams or multimedia data.

FDM uses a similar approach, except that the divisions are split between different frequencies. FDM is more common in communications networks such as those constructed for mobile telephones. Many networks use both types of multiplexing in conjunction with one another, although the most efficient use of a multiplexed channel is achieved by using statistical techniques to move data streams into time slots and frequency ranges, which are not being used at that particular moment. These statistical techniques are notoriously complex to follow, although an indication of where the next item of data can be found is usually documented within a standard agreement, or as a pointer within the actual data stream itself. This latter technique allows the communications network to make constant adjustments to improve its efficiency and, in simple terms, works by assigning a proportion of the data stream as an information block to inform the receiving device of the location of the next block of data in that stream.

1.7 Data interchange in computer networks

The subject of digital transfer cannot be comprehensively covered without an in-depth study of digital networks. Furthermore, the transfer of digital audio data over networks cannot be covered without looking at the problems that this presents. In order to understand what the issues are, so as to make informed decisions, it is necessary to understand how networks evolved into the all-pervading technology they are today.

Presented here is a very brief history of network technology, which is expanded in the following chapters where the theories are discussed in more detail.

1.7.1 A brief history of network development

Networks have developed over the last three decades or more, from the old mainframe days where multiple users accessed databases simultaneously. Occasionally services would be made available over a phone link to remote sites, other mainframes or to the public domain, at a cost. Development on networks continued during the 1970s, when some of the most significant theoretical work was completed describing the solution to interconnection (specifically the 4-layer and 7-layer models, described in more detail in the following chapters).

Ethernet
By the end of the decade the Palo Alto Research Center (PARC), run by Xerox, released the first commercial Ethernet standards. This is the same Xerox PARC that is widely credited with developing the idea of the graphical user interface (GUI pronounced 'goo-ee') that eventually became the familiar windows-based interface used by both Macintosh and PC-based microcomputers.

By the late 1980s to early 1990s, when the cost of Ethernet had dropped, companies such as Novell and others made networking more popular with purpose-built file server operating systems.

The advantage to business organizations was the ability to share the information that had become isolated as a result of the PC boom. The ability to share files and information, along with the ability to send 'e-mail' messages from one computer to another became the commercial force behind the early adoption of networks in business.

File servers

Although Apple and Atari were widely accepted microcomputer architectures for audio and print media industries, the x086/ISA combination was adopted for general use amongst commercial organizations. The architecture and microprocessor remained the same on purpose-built file servers, but by the mid-1990s these machines were developing with military-like availability in mind.

None of this really takes into account the history of the Unix operating system. Unix was (and in many ways still is) considered the 'proper' way of doing things, but its development was also affected by mergers, acquisitions, and other commercial forces. However, most of the developers and academics working on computing theory were working with one flavour or another of Unix, and it is these people who were also responsible for the development of the important network theories, which Unix itself was developed alongside. As the IBM–Intel combination grew in power and popularity, some of the flavours of the Unix operating system (OS) were also adapted to work on the Intel/IBM hardware architecture.

Client–server

The term client–server was coined to describe environments of distributed systems where software is split between centralized server tasks and client tasks. In client–server environments, a client sends requests to a server, asking for information or action and the server responds. This is analogous to a customer or client who sends an order or request on an order form to a supplier or server, who then dispatches the goods along with an invoice. In this case, the order form and invoice are part of an agreed protocol used to communicate. There may be one centralized server or several distributed ones. The client–server model allows clients and servers to be placed independently as nodes on a network, possibly using different hardware and operating systems appropriate to their function. This allows organizations to deploy relatively cheap generic clients with a fast and more expensive server.

At the time, this was a marked change from the mainframe and midrange environments in that some of the processing of data was performed on the client computer. Mainframes and other centralized environments, on the other hand, performed all the processing on the central computer, with textual results distributed to the end user. In these cases, the end user's computer would usually be a dumb terminal only capable of displaying information, with no internal computing capacity of its own.

In contrast, the client–server environment requires only that the central computer stores files and information, which are delivered to the client upon request. The client device accepts the files or information, and the central server is freed from responsibility until the next request. Client–server environments have developed since this early definition to cope with more and more of the processing, especially in transaction-based environments such as databases. In this case, the server will store the database and perform certain requests such as adding, deleting and updating records within a database, according to requests from the client, as well as performing more complex requests such as reporting and searching.

This client–server environment proved to be far more than just a trend, and has been adopted wholesale within organizations' computing environments today. Client–server is just one solution for storing and manipulating information and although it is popular, its installation does not preclude other methods, systems, and designs. It is common to see proprietary mainframes, Unix, and other systems all working side by side within the same environment.

As the client–server paradigm matured, it became possible to do more than just share files and information, until a computer program written to perform a function inside a company could also be used across the increasingly popular Internet. Due to the different computing environments on which data are stored, as well as the security risks inherent on such a freely available medium as the Internet, more complex client–server environments have developed. The flexibility of networks is such that it is possible to design client–server environments at will, and include many sources of data in a single application. Two examples of a multi-tier hierarchy are shown in Figure 1.17.

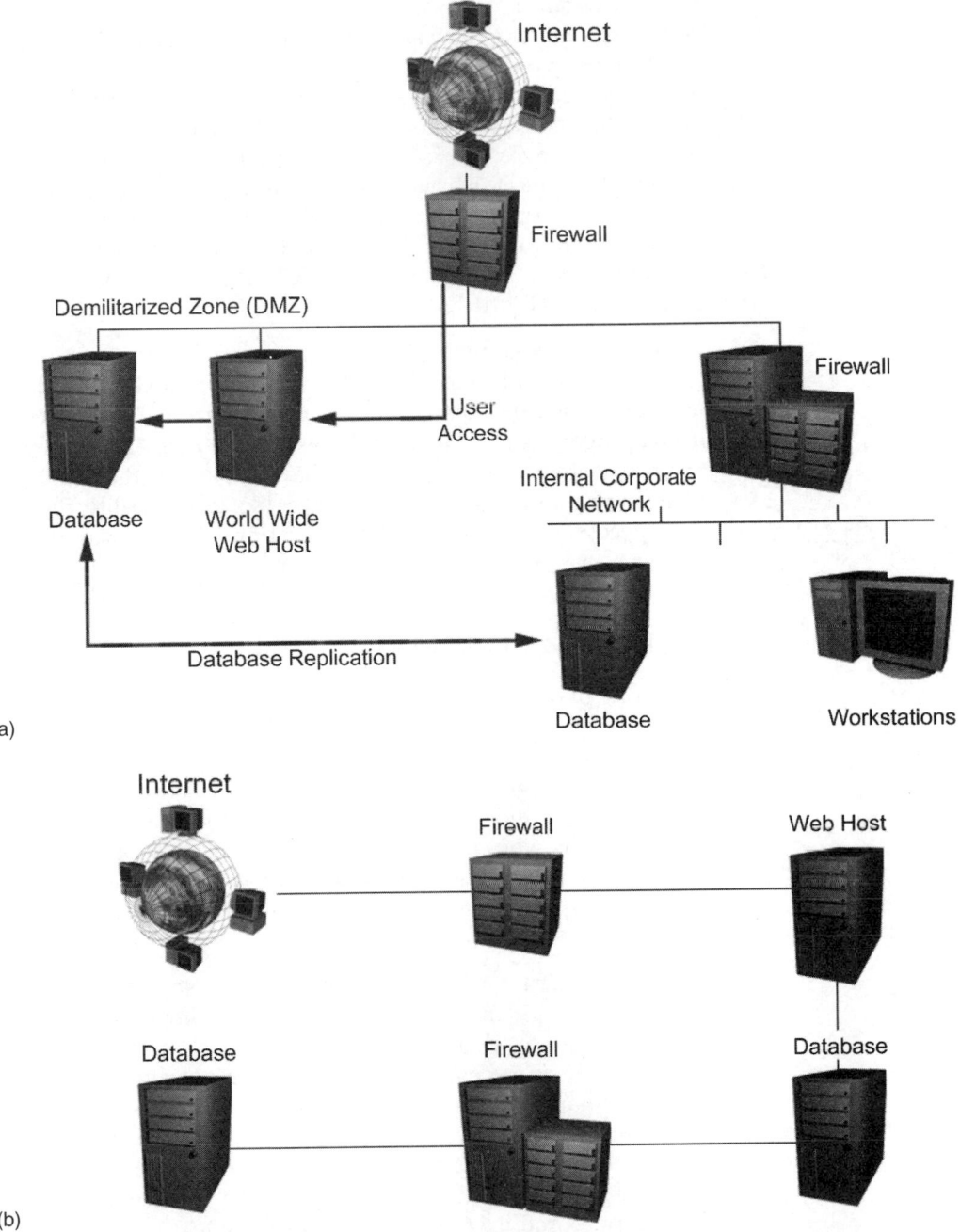

Figure 1.17 An example of multi-tier client–server environment. (a) Shows the schematic layout of computers attached to the Internet. A device called a Firewall is placed between the Internet (represented as a connected globe) and a 'demilitarized zone' (DMZ). The Firewall acts as a security device to prevent unwanted access. A further Firewall acts as a second safety precaution to protect the internal organization network. Data are replicated from the internal database server to the database attached to the demilitarized zone. (b) Shows the logical architecture of the same system.

Standardization

Although standards play an important role in the development of computer networks, in the early years as the computer industry grew more commercial, systems remained proprietary and new developments were closely guarded with each vendor protecting its own R&D investment. This created problems for organizations wanting to connect their various systems together, but the impact which the Internet has brought ironically came from free and open standards and general agreements, in the shape of transport control protocol/Internet protocol (TCP/IP) and hypertext mark-up language (HTML).

HTML is viewed through the browser software application, which was intended to allow the viewing of text-based documents in much the same way as a dumb terminal. More recent developments, such as Java, XML (extensible mark-up language) and some other visual techniques, have expanded the capability of the browser to perform client–server like functionality, fuelling the success of the Internet by creating a new way to do business.

1.8 The role of standards

Standards play an important role in digital technology, since they document agreements on how machines can be made to understand each other. Provided these languages have been defined in enough detail, manufacturers can adopt a standard, confident that their machines will talk to those of another manufacturer who has adopted the same standard. If two machines are using dissimilar languages, this will be akin to two people trying to have a conversation, when one person talks only Greek and the other Japanese!

Almost all forms of information can be stored as digital information, the only limitation being on how the original information is encoded and decoded. The most important aspect of the process is to be able to undo the original digitization and turn the binary data back into a form that is understandable, so that the analogue information appears acceptable to the consumer. In order to do this, it must be understood what measurements need to be taken in order for the process to accurately reproduce the original information.

The measurements and how they are represented are defined by standards published through international standards bodies, such as the Institute of Electrical and Electronic Engineers (IEEE), American National Standards Institute (ANSI), British Standards

Institute (BSI) and so on. Such standards are designed to communicate the methods used to encode and decode information, so that they can be correctly implemented in new systems. The normal method for writing a standard starts with some individual or organization who has a requirement for a particular standard, placing a request upon the standards organization. After investigation into the feasibility of the request, the standards organization will put together a number of interested parties – often from academic institutions and competing commercial organizations – in order to try and ratify the suggestion, and work towards the most appropriate solution to that request. Standards tend to be deeply technical, often lengthy documents which take a long time to become ratified. This can be compounded by the democratic committee-based organization of representatives from commercial or international organizations. This often has unfortunate consequences on fast-moving technology-driven industries. Work on a new standard will be announced and commercial companies will regularly start to produce equipment designed to meet the standard before it has become ratified. This can result in early implementations that do not meet the full standard and do not communicate correctly with fully compliant devices. Justifications for this behaviour are: (i) that the vendor gains experience in the standard from initial implementation; (ii) that the technology is available; and (iii) that the customer is anyway able to implement improved technology to take advantage earlier than standards are ratified.

Notes and further reading

Pradhan, Dhiraj K. (1996) *Fault-Tolerant Computer System Design*, Prentice Hall.

Rumsey, Francis (1994) *MIDI Systems and Control*, Focal Press.

Rumsey, Francis and Watkinson, John (1995) *The Digital Interface Handbook*, Focal Press.

Standard MIDI-File Format Spec. 1.1. The International MIDI Association, LA.

2 Network theory

2.1 Introduction

The previous chapter presented the fundamental principles required for understanding the digitization and transfer of audio within the context of this book. This chapter presents some of the background theory that is necessary in order to provide the framework for understanding different kinds of communication interface, and so support the choice of technology to be applied to a particular circumstance.

The main subject areas covered in this chapter are the 4-layer and 7-layer reference models and quality of service (QoS). The reference models are useful tools in understanding network technologies and the issues that they present. Both models are taught to network professionals today, and have a bearing on how network technologies have evolved. Quality of service can be interpreted as a buzz-phrase around commercial organizations and so should be treated with caution; it is a difficult phrase to define, as it can mean different things to different people, depending on the circumstance. The section on QoS puts the term in the context of computer networks, where it is a useful tool, since it puts a name to the service expected of the network, which would otherwise be difficult to describe.

The main advantage of understanding these concepts is to clarify thought, and present information that has been ratified by academic thinking and business markets. By presenting this information, it is hoped that the reader need not spend too much time conceptualizing the problems or defining the requirements of a particular subject area.

These concepts may be thought of in a similar way to project management tools. For every project manager, there is a right and wrong way to manage a project; however, it is arguable that there is no right or wrong way to manage a project, there is only the success or failure of that project to consider. There are several practices that have been proven over time to reduce the likelihood of failure, and the popularity and success of these practices means that project managers do not have to spend too much time worrying about how to best approach the organization of a complex task. Arguably, common sense, background knowledge, and experience are all important, but for the newcomer it is quicker and more useful to cross reference against practices that have already been developed and used successfully.

2.2 Bandwidth and information availability

The term bandwidth is used by electrical engineers to refer to the frequency range of an analogue signal. In the network industry, the term is sometimes used to refer to the capacity or the data rate of a link. Note that because of sampling, quantization, and compression, the bit rate needed for a given bandwidth analogue signal is potentially many times less than might be implied.

The principal endeavour of the computer networking industry is the availability of bandwidth, but no matter how high the figures achieved, there is never enough bandwidth. Amdahl's law states: one I/O bit per instruction, which roughly translates into the increasing data transport rate required as CPU capacity increases. According to the statement, the amount of data that needs to be transported by an interface is directly proportional to the rate at which the CPU can perform instructions. The amount of information created by any multimedia application (such as those which generate or manipulate audio, video or use some kind of feedback and interaction) is extremely large and sensitive to QoS, as seen in Chapter 1.

Most studios using digital audio are forced to use a combination of open-standard and proprietary techniques in order to transfer data from its source to the recording media, since one vendor may not have products that fit all the steps in the audio production chain. In such cases, interconnections, media filters, and translation devices of various types ensure translation and synchronization between the various devices in use. This method of transfer is difficult to manage and can be extremely complex, especially over long distances when all the component elements are considered.

Audio is intrinsically related to time since its perception by the ear occupies a particular time span. Digital audio data cannot usefully be heard without the correct decoding mechanism, and even then an accurate reproduction would only result if the same understanding of time is used during the playback and recording processes. Otherwise, the playback will be altered when compared to the original recording.

When audio is played back (or streamed) over networks, the real-time nature of audio becomes a problem, particularly for applications written to support real-time audio streaming over the Internet, such as Internet radio and telephony.

2.3 Data interchange in computer networks

The most appropriate way to present the issues surrounding the transfer of digital audio data over networks is to present the problems of different classes of data as they apply to professional audio use. Armed with this knowledge, it is then appropriate to discuss the capabilities of various technologies available to the studio. Most of the technologies have specific limitations in terms of the amount of data that a studio would like to be able to transfer, but others have limitations because of the role they are designed to fulfil. A more informed understanding reveals that many of the technologies have common features and mechanisms, and can be made very useful to the multimedia industry.

2.3.1 Characterization

Within information technology, a network is defined as a series of points (nodes or stations) interconnected by communication paths (www.whatis.com, 1999). Networks can be connected together to form internetworks and can also contain subnetworks.

Using such a broad definition, most normal audio facilities have several different networks already in use, although they may not be thought of as such. For example, connecting several devices together for control via MIDI creates a network, as does connecting devices together to exchange digital audio information or synchronization data.

Data communications networks can be classified in different ways. For example, a network can be characterized by its geography or by its topology. The most common topologies are bus, ring, and star configurations. MIDI is an example of a bus topology, which is formed by a single, open ended network length, whose ends do not meet. A ring configuration is formed naturally enough, by connecting the two ends of the cable together and unless the technology is designed to work under both conditions, a bus should not be turned into a ring and a ring should not be turned into a bus.

For instance, if the two ends of a MIDI network were connected together, MIDI messages might traverse the whole ring and start again, leading to all sorts of problems with timing errors and repeated messages.

Figure 2.1 shows the classic diagrammatic representation of bus, ring and star topologies that will be adopted throughout this book. Star topologies are more commonly used since the advent of hubs, which turn bus and ring topologies into a star. This is explained in more detail later.

Networks can also be characterized in terms of spatial distance (geography) as LANs, MANs and WANs, as we have seen. In audio terms, a MIDI network is usually contained within one or two rooms, and would be considered a local area network. LANs can get quite large, containing hundreds of stations, and

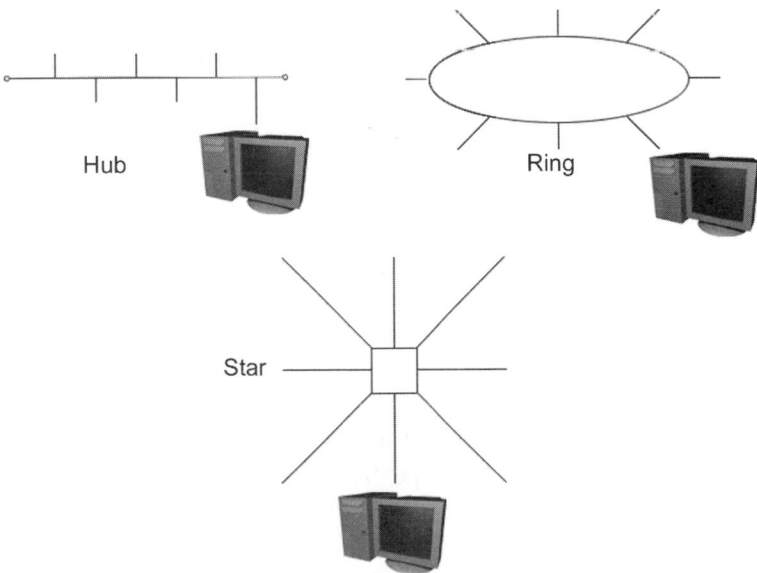

Figure 2.1 Diagrammatic representation of hub, ring, and star topologies showing a workstation attached to each.

spanning an entire campus of several buildings, or can be quite small, limited to two or more devices within a department or studio.

Networks can also be characterized by the transmission protocol being used to carry the data and this might be likened to determining whether a transport network is designed to carry cars and trucks, or railway carriages. Physical network cabling carries both types of traffic at the same time. The two different traffic types cannot see each other, although the flow of one may effect the other. This coexistence of protocols on the same cable is known as creating a logical network. A number of techniques are available for combining logical networks in this way, such as surrounding one type of data with the protocols for another, in the same way that cars may be transported on a rail network by placing them onto a carriage designed for the purpose.

Other characterizations or classifications include synchronous or asynchronous, the nature of the connections and the types of physical cable in use.

2.3.2 Packet switching networks

Packet switching networks proliferate a type of communication capable of dealing with multiple destinations, multiple devices attached to the same cable, and unanticipated connections. Although resilient to change, a delay is incurred because the amount of data being transmitted cannot be anticipated in advance. Packet switching networks are more efficient if some amount of delay is acceptable and for multiple distributed connections.

History

The US Department of Defense (DoD) initiated the early work on network technology in the early 1960s. The project to investigate the ability to maintain command and control over missiles and bombers after a nuclear attack was named as the Advanced Research Projects Agency NETwork (ARPANET) and focused on the distribution of data (Kristula, 1997). The research produced a document that described several ways to accomplish this and the final proposal was a packet switched network, described thus:

> *Packet switching is the breaking down of data into datagrams or packets that are labelled to indicate the origin and the destination of the information and the forwarding of these packets from one computer to another computer until the information arrives at its*

final destination computer. This was crucial to the realization of a computer network. If packets are lost at any given point, the message can be resent by the originator.

Paul Barran, RAND Corporation, 1962

The Internet
The ARPANET project involved both military and academic institutions and eventually the network developed into the public domain and became the Internet.

Information on the Internet architecture and related protocols are published in requests for comments (RFCs) that are available electronically via the Internet or in hard copy from the Defense Data Network DN Network Information Center (Room EJ291, SRI International, 333 Ravenswood Avenue, Menlo Park, CA 94025, USA).

Character
Packet switching networks are known as connectionless, since data travels with address information attached directly to it, in order to find its way through the network. This is explained more graphically in Figure 2.2. Furthermore, packet switching networks are often multiple access cables, with more than one device attached to the same cable.

2.3.3 Circuit switching networks

A circuit switched network is a type of communication where a dedicated channel (or circuit) is established for the duration of a transmission. The most ubiquitous circuit switching network is the telephone infrastructure, which links together wire segments to create a single unbroken line for each telephone call.

History
The history of circuit switching networks belongs within the telecommunications industry. Circuit switching networks developed to speed up the time taken to connect calls and are ideal for communications that require signals to be transmitted in real time, such as voice conversations. The popular picture of rows of girls connecting calls is an original circuit switched network.

With the advent of silicon switching, circuits can be made and broken far more quickly. The signalling bandwidth of the complete circuit can be used for the transmission of digital data. More detailed discussion on appropriate technologies can be found in later chapters.

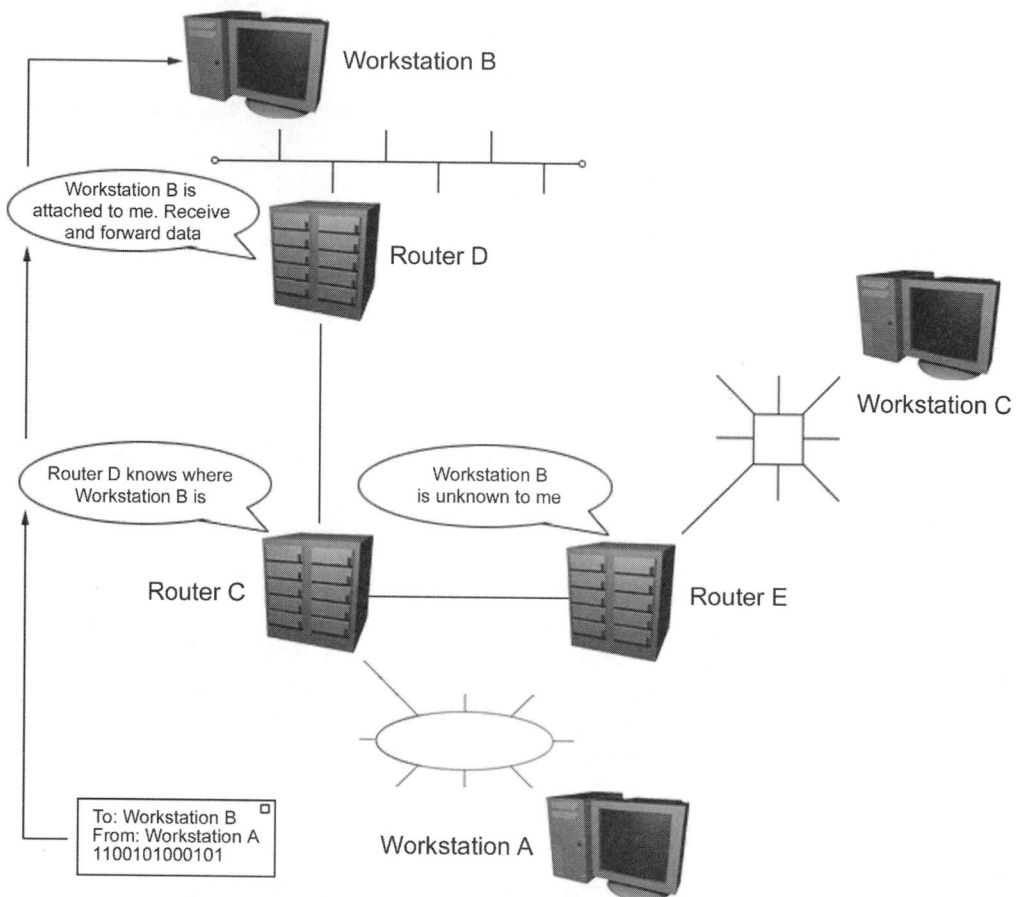

Figure 2.2 Packet switching networks are commonly multiple access in nature, meaning many devices connect to the same cable. Data sent over such a system require addressing information to determine the sender and recipient. Devices called routers hold a list of attached devices in tables known as routing tables. This allows information regarding attached devices to pass throughout the network. In the case that a device is unreachable, the packet will be dropped from the network. In the case illustrated, workstation A requires to send a message to workstation B. This might be within the same building, or across the Internet, the principle is the same. The nearest router sends out a message requesting if any other attached routers know about workstation B. Router E responds with a message indicating that no match was found within the its routing table, whilst router C receives a response from router D indicating that a route to the destination has been found.

The infrastructure was developed to transmit data over longer distances, and the appropriate acronym POTS stands for the Plain Old Telephone System. Appropriate progress has seen circuit switching technologies develop into a more ubiquitous technology, available for use more generally.

Character

Circuit switching networks are sometimes called connection-oriented networks, meaning that a connection appears to have

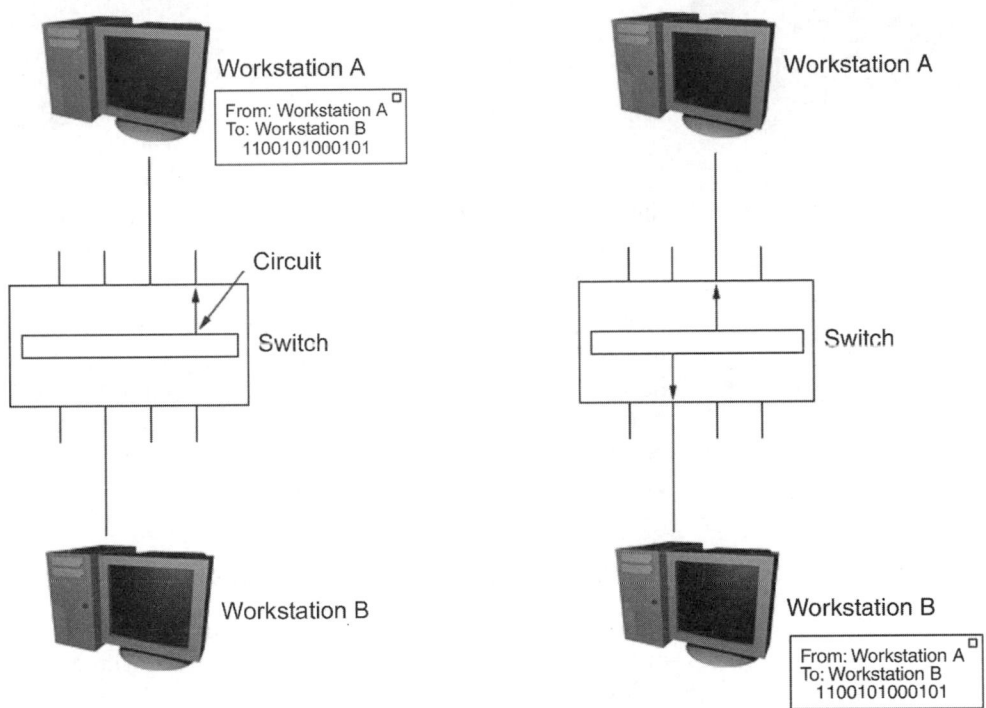

Stage 1:   Stage 2:

Figure 2.3 Circuit switching. Stage 1: workstation A sends data to workstation B. Stage 2: to expedite the transmission the switch creates a physical connection between the sending and receiving device for the duration of the call. With the advent of fast switching equipment, the time to set up a call is becoming negligent, with the result that shared media technologies (such as Ethernet) benefit from switched circuit technology (in the case of Gigabit Ethernet).

been made directly between the two devices at each end of the circuit (Figure 2.3) and should not be confused with connection orientation mentioned further up the 7-layer model. Packet switching networks, whilst being connectionless, are made to be connection oriented by using a higher-level protocol. Transmission Control Protocol, for example, makes Internet Protocol networks connection oriented.

2.3.4 Point to point connections

Point to point connections might not present themselves comfortably as a network type, but the use is generally to connect devices together for the purposes of communication or transfer of data.

A point to point connection would normally be associated with two devices, one connected to another, such as for copying a

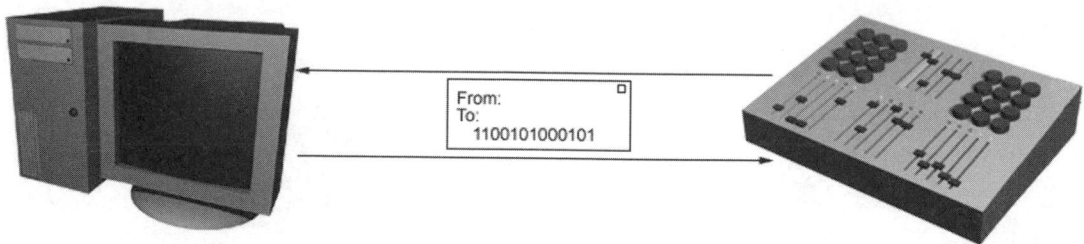

Figure 2.4 Point to point connections are generally described as connecting devices directly, as shown. In such a system, addressing information is redundant and not included, since there is only one other device from which the information can be sent.

digital audio tape from one machine connected directly to another, as shown in Figure 2.4. The connection is enclosed, and no other devices can see or address the network.

Although most common audio standards utilize a point to point topology, connections can be more complex over other physical network types by creating a logical connection between devices. Transmission Control Protocol for instance makes connectionless networks connection oriented, whereas Fibre Channel takes a different approach to multiple access.

2.4 US Department of Defense 4-layer model

Among the many significant publications from the ARPANET project was the Department of Defense 4-layer architecture (4-layer model), illustrated in Figure 2.5. It proved difficult to design protocols adhering to this model and it was several years before the transport control protocol/Internet protocol (TCP/IP) was finalized and deployed.

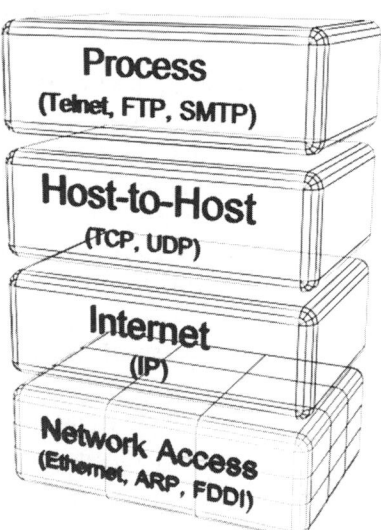

Figure 2.5 US Department of Defense 4-layer interconnection model, with examples of protocols in each layer, for later reference.

In the meantime, the International Standards Organization worked on the 7-layer model, which identifies the steps that need to be addressed for general communication purposes.

Although this book uses the 7-layer model as a reference, the 4-layer model is also important and is explained briefly under the following four headings.

2.4.1 Network access layer

The network access layer is responsible for delivering data over the physical cable and can be considered to be attached to the actual physical medium in use, such as copper or fibre optic cable.

The network access layer defines the resistance, voltages, and access mechanisms in use. Definitions exist for the restrictions in cable lengths, appropriate cable types, number of connections and so on. This layer defines the restrictions in terms of the network or single cable segment and describes how differences in voltage (or light intensity for fibre optics etc.) are turned into recognizable data.

2.4.2 The Internet layer

The Internet layer is responsible for managing the delivery of data across a series of different physical networks that might interconnect a source and destination machine. Routing protocols are associated with this layer, such as the Internet Protocol or IP, the Internet's fundamental protocol. This is the most significant layer when considering connecting networks together to create an internetwork, since it describes how data are sent over longer and more complex routes than can be defined by the physical layer.

2.4.3 The host to host layer

The host to host layer deals with problems once the data have made their way from one machine to another, such as handshaking between computers. This is a common technique even in audio machines whereby connections are negotiated in much the same way as DAT machines, modems or fax machines might do with frequency or bit rate.

In more detail, handshaking is the exchange of information between two devices in order to establish the protocol to use for communication (Figure 2.6). Since the devices at each end of the

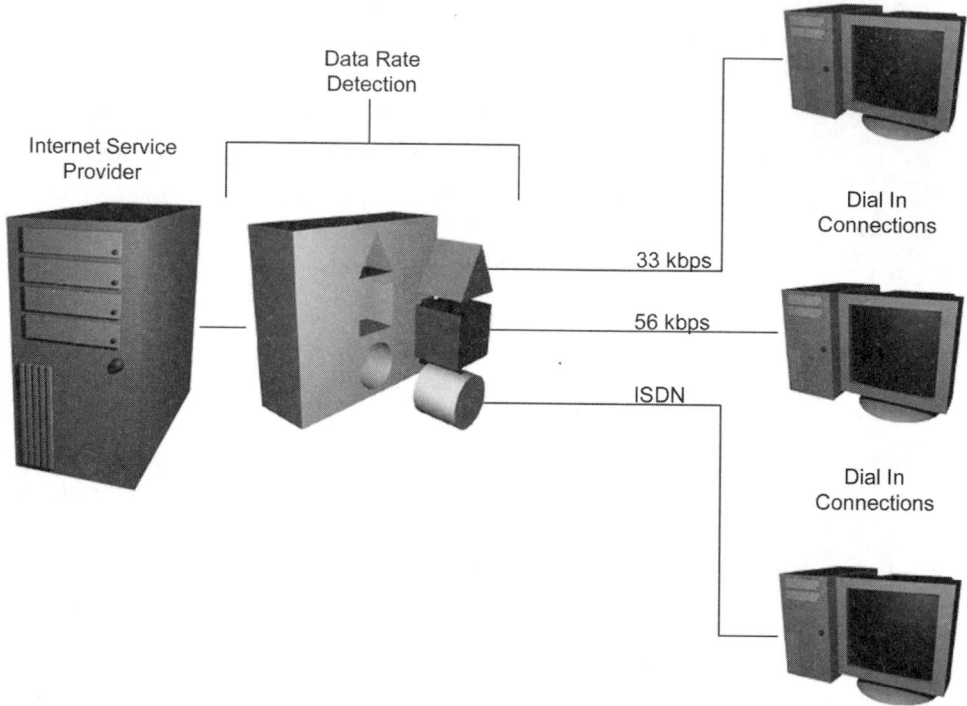

Figure 2.6 Handshaking is the key to successful communication as it is the process by which devices negotiate communications' capabilities. A single device can have multiple capabilities, and the handshaking procedure will generally begin by trying to communicate at the latest revision of a capability. For instance, in this diagram, data rate is represented. The modem receiving the call will sound a tone to indicate the best data rate at which it can converse. The dialling modem will respond to the tone only if it is also capable of connecting at that data rate. If not, the receiving modem will sound a new tone to indicate a slower data rate. This continues until a match is found, or all available rates have been exhausted.

line may have different capabilities, a handshaking procedure is used in order to determine the highest transmission speed that both can use. In DAT machines, handshaking is used in order to determine the sample frequency and word length of the incoming signal for instance, much like establishing the correct shaped key in order to open the door of further communication.

2.4.4 The process layer

The process layer contains protocols that implement user-level functions, such as mail delivery (SMTP – simple mail transfer protocol), file transfer (FTP) and remote login (TelNet). Each of these provides a different type of service over the network.

Further details of the 4-layer model are incorporated in descriptions of the 7-layer model in the next section.

2.5 The ISO 7-layer open systems interconnection reference model

Taking the lead from the 4-layer model, the International Standards Organization (ISO) undertook a redefinition of the model in order to address what was seen as the shortfalls of the original. The result, which never became a ratified standard, was the ISO OSI reference model (see Figure 2.7), which is taught to network professionals today and known hereafter as the 7-layer model. Protocols adhering strictly to the OSI model are notoriously difficult to understand and are indeed bloated and inefficient (Entry for OSI Seven Layer Model, 1999), but its most significant contribution is the underlying philosophy of networking as represented by its layered model.

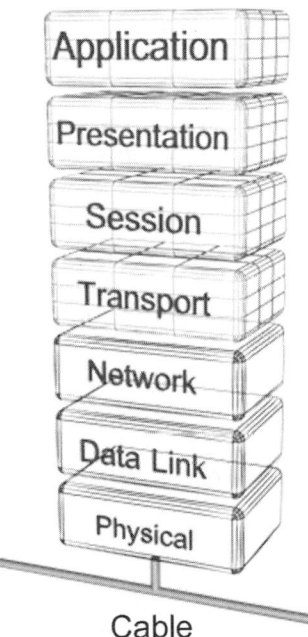

Figure 2.7 The International Standards Organization Open Systems Interconnection 7-Layer reference model (referred to as the 7-layer model within the text).

The model describes, in seven layers, all of the problems that need to be solved in order to allow multiple computers to talk to each other on a network.

It is useful to understand the model and what it presents, since the descriptions of the processes contained within the seven layers describe all the steps necessary to place data on a cable for receipt by a known recipient. The modularity of the 7-layer model allows it to be a useful tool wherever it is necessary to place data on a cable (or other transport medium). It is useful to think of digital audio transfer in these terms, as the exercise illustrates what steps have been implemented and therefore what functionality is missing from the candidate solutions and architectures.

Although the 7-layer model is considered to be an improvement over the older DoD 4-layer model, one of the layers was subsequently split into two further sub-layers.

The 7-layer model's basic functions match those of the 4-layer model, although the work of the ISO/OSI committee described the functions in more detail, leading to the increase in the numbers of layers. It is easy to get bogged down in the theory behind the 7-layer model, and it is worth taking a step back to look at the work as a whole. By doing so, it is easier to appreciate that it describes every stage in the translation of electrical signals into comprehensive and error-free data communications between digital devices. As with the 4-layer model, the OSI model is designed to be modular. That is, any mechanism or set of rules designed to fulfil the requirements of a layer can be interchanged with any suitable mechanism for that particular layer, without affecting the functional mechanisms applied to the adjacent layers. For instance, two mechanisms that fulfil the requirements of the physical and data link layers are token passing and CSMA/CD (or Ethernet and its cousins). Each has very different rules of engagement, and may even utilize different cable types or electrical parameters. However, this should not effect the ability to run upper layer protocols, such as TCP/IP upon the physical medium. This would give rise to common descriptions such as 'TCP/IP over Ethernet' which describes several functional layers of the model.

As with the 4-layer model, it remains difficult to implement the 7-layer model with exact partitions at the layer boundaries, and a good deal of energy is expended by the network industry in trying to get new network technologies to talk to established ones. The boundaries between the layers would be better illustrated with dotted adjoining lines, since functionality often affects adjacent layers.

The table shown in Figure 2.8 shows the functions and methods associated with each layer. To understand the model, imagine that the network cable is attached to the bottom of the physical layer, as suggested in Figure 2.7.

OSI layer	Topics	Methods
Application	Network services	(See Section 2)
	Service advertisement	Active
		Passive
	Service use	OS call interception
		Remote operation
		Collaborative
Presentation	Translation	Bit order
		Byte order
		Character code
		File syntax
	Encryption	Public keys
		Private key
Session	Dialogue control	Simplex
		Full-duplex
		Half-duplex
	Session administration	Connection establishment
		Data transfer
		Connection release
Transport	Address/name	Service-requestor-initiated
	Resolution	Service-provider-initiated
	Addressing	Connection identifier
		Transaction identifier
	Segment development	Division and combination
	Connection services	Segment sequencing
		Error control
		End-to-end flow control
Network	Addressing	Logical network
		Service
	Switching	Packet
		Message
		Circuit
	Route discovery	Distance vector
		Link-state
	Route selection	Static
		Dynamic
	Connection services	Network-layer flow control
		Error control
		Packet sequence control
	Gateway services	Network layer translation
Data Link–LLC	Transmission synchronization	Asynchronous
		Synchronous
		Isochronous
	Connection services	LLC-level flow control
		Error control
Data Link–MAC	Logical topology	Bus
		Ring
	Media access	Contention
		Token passing
		Polling
	Addressing	Physical device
Physical	Connection types	Point-to-point
		Multipoint
	Physical topology	Bus
		Ring
		Star
		Mesh
		Cellular
	Digital signalling	Current state
		State transition
	Analogue signalling	Current state
		State transition
	Bit synchronization	Asynchronous
		Synchronous
	Bandwidth use	Broadband
		Baseband
	Multiplexing	Frequency-Division (FDM)
		Statistical
		Time-Division (StateTDM)

Figure 2.8 Table showing functions and layers associated with the ISO/OSI model.

2.5.1 The physical layer

The physical layer provides the services associated with physically connecting to the cable and creating the link. These include the encoding of data into voltages, or whatever binary states the medium requires (such as frequency in fibre optic cable, microwave, and infrared linkages) and the encoding of the data using mechanisms such as non-return to zero. This layer is responsible for physically transferring messages between nodes. The physical connection may take several forms, and the path of a message through an internetwork may use several different types of physical layer mechanism before reaching its destination.

In simple terms, the physical layer can be thought of as attaching to the cable (or other medium) and being responsible for identifying data upon it. The physical layer is also responsible for configuring the link. This means that various other parameters are defined here, such as the maximum number of stations on a segment, the voltage and impedance used on the medium, the maximum length of cable and so on.

The physical layer provides the data link layer with a method of moving bits between two machines. It is recognized as error prone and unreliable because of the conditions, such as geographical distance and environmental changes including radio frequency interference, through which the physical layer carries information. As a result, other layers ensure data integrity by using error checking and correction mechanisms.

2.5.2 The data link layer

The data link layer provides the error checking functionality that makes the unreliable physical layer reliable. In addition, the data link layer also provides flow control such that messages of a given size may be reliably transmitted.

This layer has subsequently been split into two sub-layers. The sub-layers are called the media access sub-layer (MAC) and the logical link control sub-layer (LLC). The MAC sub-layer is responsible for those functions that are associated with the physical layer directly below it, such as the error control that is required in order to make the physical layer reliable. This will normally include some mechanism whereby a packet or frame, as it is called at this layer, can be resent if it is lost or damaged, as well as determining whether the cable is available to accept transmission.

The LLC layer on the other hand concentrates on taking data from the layers above and marshalling them downwards, and so has more to do with controlling the flow of the data, and splitting it into correctly sized messages (packets, frames or datagrams) for transmission.

2.5.3 The network layer

The network layer is an important layer for connecting networks together to form internetworks, since it is responsible for routing messages between nodes. When a message reaches an intermediate node in the network, the network layer implementation at that node will be able to determine which route that message should take on the next stage of its journey to the destination device. This may entail determining which link, of several connected to the node, should carry the message onwards. In order to operate efficiently, the network layer uses its own information gathering systems called routing protocols, and several of these are in use, such as those covered in the next chapter.

Although network layer protocols tend to exist between nodes, rather than at the sending and receiving stations as shown in Figure 2.9, they do effect the original data, by altering addressing information, which allows other nodes to determine the destination of the data. In this way, devices operating specifically at the network layer are able to determine the correct route or path that the packet needs to take in order to find its destination.

Figure 2.9 Network layer protocols live within the network and are not directly related to the communication of two end stations, but rather manage the connection. Different device types are associated with different layers, and communicate with other devices at the same layer. For instance, routers are so called since they make decisions regarding the appropriate route for a particular message (or packet) depending upon its network layer address.

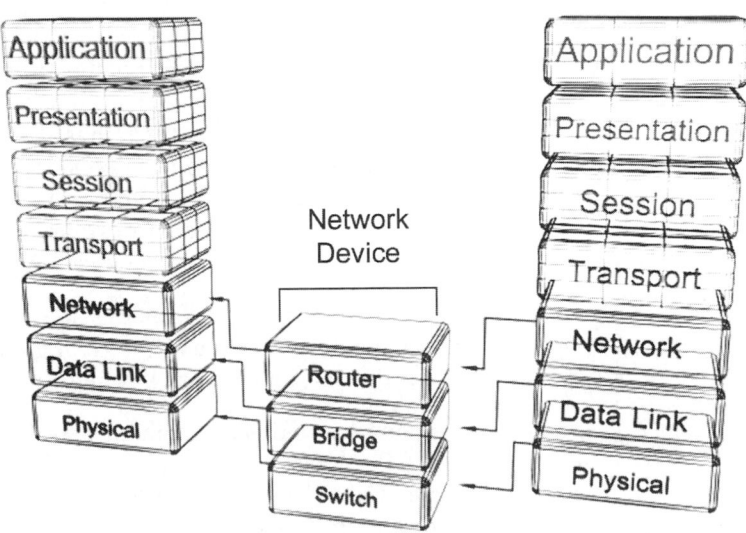

Each device in the network layer may strip the original addressing information from the frame and add its addressing information, or it may append more addressing information to the frame, making it longer. The disadvantage of appending addresses is obvious – frames get longer. Some devices operating at the network layer may have a maximum frame size, and so frames that are too long may get split into the correct size until eventually, the original data form only a very small part of the information that arrives at the destination. It is very important to retain this information if the returned data are to navigate their way back to the sender. On the return journey, each network layer node strips the new information away until the data are in the intended form at the time it reaches the original sender.

Network layer devices have their own sets of protocols and rules, which are used to communicate the status of particular links to one another in order that the fastest routes for data to travel may be discovered before sending the data onward.

Taken together, the first three layers are referred to as the subnet layers. To provide full connectivity, all subnet layers must be implemented in each node of the network

2.5.4 The transport layer

Layers 4 through 7 have less to do with the mechanisms for moving information around, since at this point the data have been created and moved from one node to another. These layers are collectively called the end-to-end layers because their services are required only in the end nodes and not in the intermediate nodes. The transport layer manages end-node to end-node communication, delivering the units of data of whatever size, from one device across the network to the receiving device. While the data link layer ensures that a message will not be damaged, the subnet layers may not necessarily guarantee that all messages will be delivered, or if they are delivered, in what order. The transport layer is therefore called upon to handle flow control, retransmission and message sequencing.

The transport layer provides long-range resilience to the data. Since data can arrive in the wrong order, the transport layer will reorder the data before passing to the upper layers.

Examples of this mechanism appear in transport control protocol (TCP), where each packet of data is numbered so that the transport layer can restore the sequence of packets at the receiving node.

The transport layer also inspects the number of each packet, to ensure that each one has turned up. In the event of a packet being lost, the transport layer requests that the packet is resent.

2.5.5 The session layer

The session layer is responsible for establishing, managing, synchronizing and terminating network sessions between devices. This allows the host computer to assign resources to conversations between itself and the sending station.

The session layer mechanisms take the form of handshaking, and negotiating the format and rules for the conversation to take place. Once this has been set up, there is little activity at this layer until the conversation is disconnected, whereupon the session layer will handle the termination of the session in a controlled fashion, so as to ensure that the resources become available for further conversations.

This layer is especially important in the Internet, where popular Internet sites may receive millions of requests per hour, all needing to be sorted and allocated resources in the most efficient way.

Each frame of data contains the resource allocation identifier, so that the incoming data can be immediately identified and routed to the correct resource within the host device.

2.5.6 The presentation layer

The presentation layer takes care of translating data to and from standardized formats, as may be negotiated by the users. This may include character set translation, or any of several forms of data conversion, for example between two sets of floating point numbers. While layer 6 transformations often serve to present data to the receiving application in a convenient and mutually understandable format, they are also carried out for purposes directly related to the communication, such as data encryption, compression and so on.

2.5.7 The application layer

By the time the data reach the application layer, they are now in the form that the receiving program can understand, and have been stripped of all the data that may have been added to it for the purposes of the network layer (and all the other layers). Furthermore, any compression or encryption has been removed and the data are available for direct processing.

The application layer forms an interface to the user program and is often understood to encompass an application program that performs a network specific user service, such as file transfer or mail delivery.

Digital audio data could be thought of as sitting within this layer, since it only needs to be decoded by the D/A conversion process to be understood. More strictly, the D/A process itself would be one of the services offered by the application layer.

2.6 Quality of service

It is the intention of this book to offer best-practice information on the implementation of networks for audio purposes, only where appropriate. This information is referenced where necessary but conflicting information can often be found. Best practice is too often dependent upon the restrictions that are imposed upon any particular project. For instance, it may be desirable to install the best possible network technology within an establishment, but this is impossible because of the budget assigned to the project. On the other hand, it may be desirable to install the cheapest or most commonly used network technology, but this is useless for the purpose for which the network is intended.

The latest advances in communication technologies have made it possible to develop new services, particularly multimedia services such as audio and video retrieval and videoconferencing. The wide deployment that the Internet has recently experienced has raised interest in providing such services over networks in general. However, the main problem that these services present is that their requirements are completely different from those of traditional data communication. To transmit audio or video flows it is essential to preserve their time dependencies, but most current networks do not guarantee the data delivery within any given amount of time.

2.6.1 What is QoS?

A great number of parameters have to be understood by both the vendor and consumer before the correct network design can be realized. Other than cost, installation times, network management facilities, failure recovery, and traffic throughput, there is another set of parameters that make up the quality of service (QoS) that is demanded from the network. As mentioned previously, different types of data have different delivery requirements, and these help to determine the QoS.

The term 'quality of service' is more accurately expanded to, 'the quality with which a service is offered', and can be applied to any service, from car washing to network provision. To understand the term in its appropriate spirit, the definition found in *Webster's Dictionary* under the entry for quality assurance reads in part:

> *an evaluation of the various aspects of a project, service or facility to ensure that standards of quality are being met.*

The service in question is the ability of a network to transfer data to agreed requirements. A suitable network will be one whose QoS is at least as good as those required by the data type.

2.6.2 Defining the parameters of QoS in digital audio transfer

Defining the quality of service has always been an important part of specifying a network, since it would be bad practice to install a network that struggled with e-mail, when it was installed to deliver real-time audio, although the requirement was not always associated with the phrase. There is no technology that can deliver unlimited network bandwidth, on demand for all the possible types of service. Since this is the case, it is necessary to understand what sort of network will fulfil the requirements of a particular environment. A full analysis of the requirements will lead the network designer to choose the correct technology, and should have a significant impact on the eventual design of the network. It is the job of the network designer to get enough information about the requirements in order to ascertain which technology and design fulfils the requirements.

To illustrate this by example, a small studio whose customers require soundtracks for multimedia training and games is running a database library of reference sounds, short audio clips, speech, and music. Although there is a need for recording audio onto hard disk, the studio manager agrees that the recording activity and processing of files in a multi-track environment can occur at the DAW located in the studio area as shown in Figure 2.10. The server containing the database is located in a purpose-built machine room elsewhere in the facility. The network is required to deliver audio files to the workstation when a project is opened from within the DAW software. At the point of opening the project, the software requests the associated project files from the server and several hundred megabytes of data may be sent over the network from the server to the workstation. Once these have been received, the files are loaded onto the workstation and the network is no longer required.

Figure 2.10 Example Studio. In this small studio example, audio files are stored in a central database located in the machine room, but audio is processed locally using digital signal processors attached to the digital audio workstation.

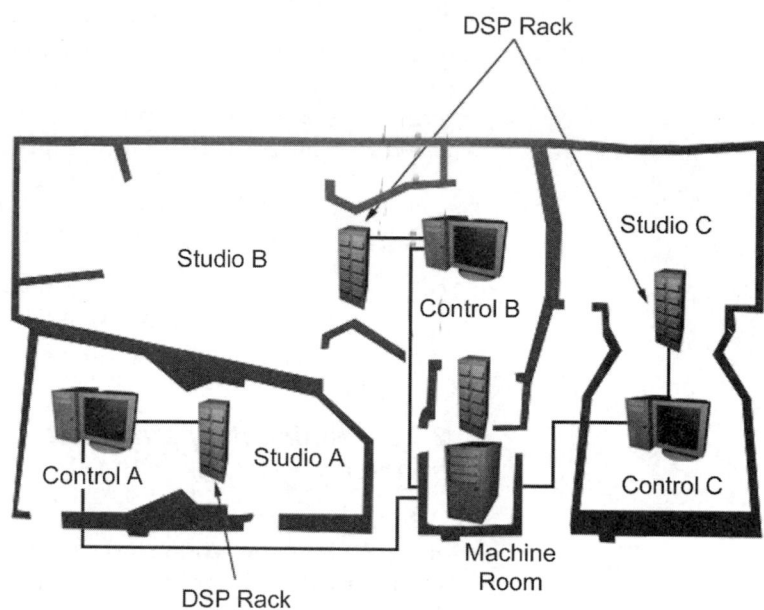

With planned and tested configuration, the workstation could be disconnected from the network and the engineer could continue normal operation. This offers some resilience in design, and ensures that a network failure would not prevent the audio engineers from performing their tasks or cause them to lose any information in the middle of a production session. The studio manager is keen on implementing a networked solution so that the DAW files are consolidated in one place for safely backing up, and can be made available to other studios within the facility. In this example, the studio manager is also keen to see the system put in place, so that the disk space can be monitored, and studio charges calculated by factoring the engineers' time together with the disk space used.

The eventual installation must not compromise the engineers' productivity but should enhance it. The files should therefore be delivered in a timely fashion so that the engineer is not kept waiting too long, each time a project is loaded, as this time is unproductive and soon mounts up. In order to fulfil the parameters laid out by the studio manager, a packet switching solution may be considered, since although the time factor has been specifically mentioned, it is not as important to the 'file-loading' nature of the description as the cost.

In the next example, illustrated in Figure 2.11, the studio manager of a large post-production complex has specified the requirements.

(a)

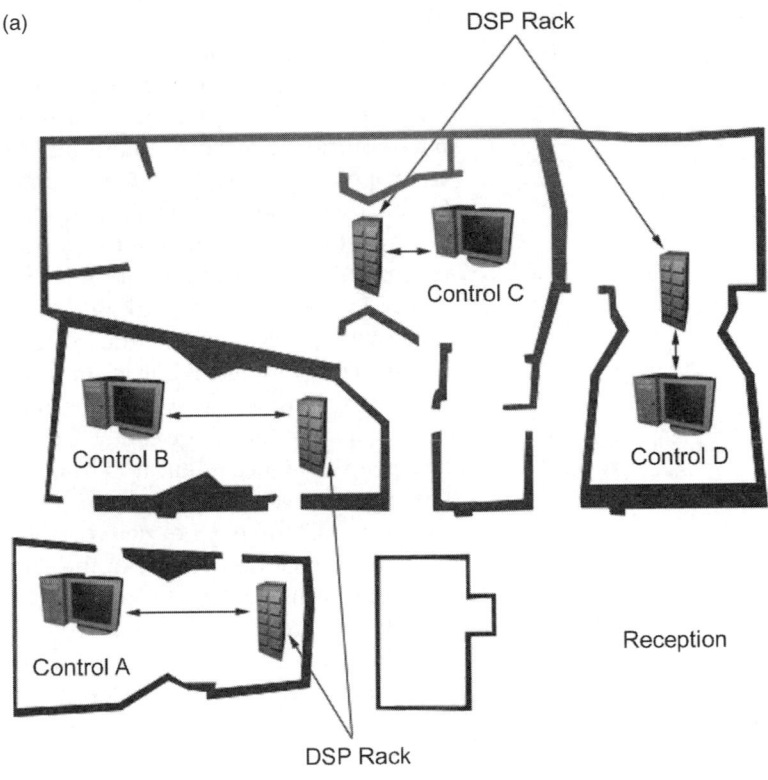

DSP Rack

Control C

Control D

Control B

Control A

DSP Rack

Reception

(b)

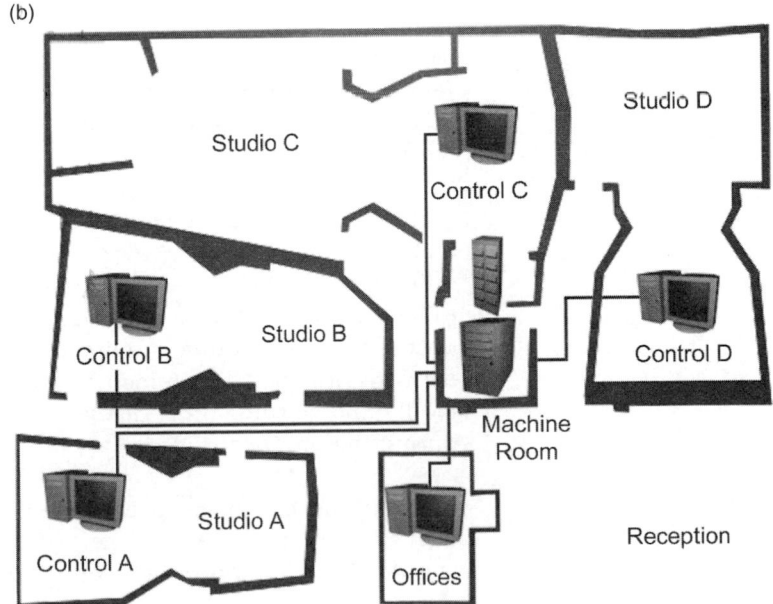

Studio D

Studio C

Control C

Control B

Studio B

Control D

Machine
Room

Control A

Studio A

Offices

Reception

Figure 2.11 Example Studio. (a) This larger studio has isolated resources, unable to talk to each other. (b) The new network installation connects each studio to the machine room, where the DSP rack is displaced, and audio files and projects are stored. In this way, each studio can use resources and audio files from the other studios.

The complex has a centralized machine room and racks full of signal processors located in each studio. The studio manager would like to take the rack-based digital signal processors (DSP) out of the studios, placing them in the secure and environmentally controlled machine room; all the signal processors can be controlled remotely. The idea in this example is that the DSP is bunched together as a single resource and is available to any one of the studios in the complex and so can be treated as a commodity. The only limitation that the customer is prepared to accept is the limitation of the power of the DSP resource, so the network must not become the bottleneck. The studio manager is concerned that equipment is underutilized and wants to find a way of dishing out DSP on demand to any of the studios that need it, so that the usage can be measured and resources planned and justified. Once again it is important that the productivity of the engineers is not compromised in any way, meaning that the audio must travel the network from the studio, enter the processing rack and return to the studio for monitoring without any significant delay. In order to achieve this, the network designer has identified that a streaming solution is required, but that the network must be segmented in such a way that the best performance can be administered. The answer in this particular case is to ensure that the delay is incurred only after the recording has been completed and therefore to record audio onto a local workstation. Once the recording has been made, a slight delay, in the order of less than 5 ms, may be acceptable only during playback. In this way, the engineer decides on the DSP algorithms (for instance equalization, reverberation, compression, and so on) and sends the control information out to the DSP resource rack, which configures the DSP resource. The resulting audio experience can be compared with the original unaffected recording.

This scenario reflects the normal studio modus operandi, except that the audio never leaves the digital domain. The only difference is that the cable topology allows processing equipment to be located further away making it more flexible in terms of operation. The solution presented is only possible because the network designer determined that audio was initially recorded locally and so some time–displacement delay could be incurred during the production phase.

Even so, the network capacity needs to be significantly higher than the first example, and the possibility of delivery failure of any data was not an option once the audio stream had begun transmission.

The two examples presented here highlight the differences in requirements between two studios. In the first, a file store-and-forward solution will suffice, whilst the second specification requires a streaming solution. These two requirements would very rarely lead to the installation of the same network technology, and it is these differences that are expressed in the definition of the quality of service of each of the networks. The QoS is determined by the functional requirements that the network is expected to fulfil, and by the budget available for the project.

There are many parameters to bear in mind when designing a network, such as the physical location and type of the cables, any security, failure recovery and redundancy that may be required. Even the details of the storage throughput and the service record of third party providers (in the case of managed services such as those provided by telecommunication providers) might need to be considered.

2.6.3 Determining the QoS

Before looking at the different types of data networks, it is necessary to understand the demands and solutions that any proposed network is expected to undertake. Because unlimited bandwidth is not available to unlimited users, it is necessary to determine the exact nature of the services that the network is intended to support. As much relevant detail as possible should be acquired before attempting the design.

It would be a relatively simple task to determine the QoS necessary from any mechanism employed to transfer audio in the examples above. The first example could be satisfied with a simple LAN technology, such as fast Ethernet (100 MB/s) or Token Ring (16 MB/s or 32 MB/s). The second would benefit from a more rigorous design, perhaps utilizing a technology such as asynchronous transfer mode (ATM) or Fibre Channel.

The second example is perhaps the most interesting from the point of view of most audio facilities, and part of the QoS can be determined using a fixed calculation. The studio has the requirement to move a number of streams of audio data around, and is using hi-resolution audio within the facility. The amount of data can therefore be calculated as

96 000 (no. samples per s)
    $\times$ 24 (word length in bits which will store each sample)
    $\times n$ (no. streams)
    = 2 304 000 bits/s per stream

From the above calculation, it can be said that there must be a data rate of around 2.4 Mbits/s guaranteed for each stream on the network. Furthermore, since the requirements of the professional audio industry are being discussed, it is critical that the integrity of the data remains intact, and so delivery must be perfect every time.

Perhaps the most difficult part of determining the strategy from the point of view of the network designer is to determine the number of audio streams that might be in use at any one time on the network. The number of streams in the audio industry varies, and the acceptable number increases regularly.

Audio has not yet made a wholesale switch to using network technologies. This may be because there are relatively few uses to which limited bandwidth can be put in audio terms, because of the amount of data that digital audio create and the limitations that audio transmission places on the network in terms of time. This is compounded by the nature of public networks such as the Internet, which provide packet-based delivery, unrelated to time in the way that audio playback demands. However, this is not to say that audio has been put to all the possible uses for the current bandwidth limitations.

2.6.4 Identification of usefulness

Some of the uses for conveniently available audio are listed below. This is not meant to be a complete list, and includes common uses for networks in the audio industry that are already in use, those in development and some which are intended purely to spark the imagination.

Library management (databases of sounds)
Automatic recall of sounds from hotkeys (such as radio jingles)
Attributes and information stored along with the audio
Transfer of control of audio equipment
Real-time manipulation
Video dubbing and multimedia
Language labs and translation
Telephone and videoconferencing
Speech recognition
Voice control (of appliances and computers)
Audio analysis
Recording and playback
Media conversion

In terms of QoS, it is the availability of the audio as a time-displaced service that stands out as common to most of the items listed above, and it is this which separates the requirements of the audio industry from such distribution networks as broadcast and retail.

2.6.5 Broadcast

In traditional broadcast networks, such as those employed by television and radio, once the broadcast of a particular item has begun, any delay that occurs once the transmission has started will spoil the experience of the recipient. If a consumer misses a broadcast, then the experience will not be repeatable in any way other than employing some method of time displacement.

Within the audio industry, it is the availability and speed of delivery that make the digital format useful, or not. For example, if an audio engineer in a particular studio required the availability of an item that was only available once, as in the public broadcast example, then this method of distribution is completely useless (unless some other time-displacement method was used to capture the broadcast, as mentioned). If the same engineer waits for a CD to be delivered through a retail distribution network, then this is also less than ideal, because of the excessive period of time spent waiting for the material, during which the engineer is redundant.

2.6.6 Categories of QoS

By narrowing the analysis to professional use only, the extremes can be excluded and it is possible to categorize the remaining requirements into four categories of quality of service:

User oriented

User oriented transmission offers subjective image and audio streaming quality, but is capable of file transmission as per the first of the studio examples used above. File transmission has a low time dependency, since the files are being delivered from one place to another before being manipulated. However, the requirement for reliable transmission remains high, so that any file remains intact following delivery. Although streaming is possible, quality is not guaranteed over time and any applications attempting to utilize streaming must take care to account for delivery failure of individual packets, and structure the stream to an inconsistent QoS.

In general, a single file is requested from a device attached to the network, and this is delivered to the client computer, where it is loaded into memory for processing by software control. Since time is not a significant factor in this situation, it is possible to increase the reliability of an otherwise unreliable mechanism by retransmitting any lost information, provided the file is reconstructed afterwards, in order to satisfy the requirements.

Consumer broadcast

Although consumer broadcast in its traditional sense will not be covered specifically, this general heading also covers on-demand consumption, and one-to-many transmissions.

Internet radio broadcast is an example worth looking at, where a consumer may be searching for a service that covers a particular event, such as a sporting event not covered by traditional broadcast media.

The consumer carries out a search for an Internet service providing coverage of an event. The Internet service provider (ISP) will provide a commentary on the event, which is turned into a digital stream. The event is not 'broadcast' in the traditional sense, since the information will only be sent to computers which have specifically requested the stream. A slight delay of a few seconds between the transmission and receipt will not affect the enjoyment by the consumer, and may be an anticipated part of the service. In fact such a delay is common even in television broadcasts, where an event may be taking place on the other side of the world, and a propagation delay is incurred because of the distances involved.

For the purposes of the consumer, this is still considered 'live', although for live musicians attempting to play together over the link, the delay makes the task impossible to achieve. This was demonstrated during the famous Live Aid broadcast in 1986 where such an attempt resulted in an impossible time-keeping task for the performers.

In the case of Internet radio's coverage of sporting events, the transmission may be the only coverage of the event, and so the consumer will be more inclined to accept poorer quality reception, involving dropouts and compressed audio. This is not to say that it is acceptable to provide information that is so degraded as to impair important information about the event. In other words, there is a point when the service becomes so degraded that the consumer will not accept the service. Therefore, the clarity and continuity of the final output has a level that can be defined. Work in this area is where Internet broadcast software manufacturers do much of their research.

Another common definition of on-demand QoS that is often associated with delivery of video is that the consumer can choose when to watch a particular program. An example of this, which is discussed in the book by Bill Gates (CEO of Microsoft Corporation at the time) entitled *The Road Ahead* is that a popular film is made available by a service provider for a period of time, say a month. During this time, the viewer can ask to see the film at a time that is convenient to him or her. At the time requested, the film is broadcast to that viewer, and any other viewers who have also requested the film at that time. This scenario is some way from being reality for most people, although early tests have been carried out and the service is being operated in a few controlled locations (Lucent Technologies, 2000).

Format and synchronization oriented

Format and synchronization oriented QoS includes video resolution, frame rate, storage format, compression schemes and the skew between the beginning of audio and video sequences.

Commonly associated with studio processing this service level is the first of the uses for audio transfer that increases the QoS demands towards specifically designed networks. In this category, the QoS can suffer from a small delay such as that defined in the second option, consumer broadcast, above, but there must be no deterioration in the quality of the audio that is received. Applications for this would be in almost any typical studio where an audio signal is being engineered through processors and analysers before being submitted as a final product. Specifically, this may involve streams of audio being sent to processors such as mixers, equalization, effect units, and audio compressors.

In order to retain the highest possible quality of the audio, there must be no dropouts in the stream. The most obvious example of this is when streaming to a tape mechanism, when a dropout in the stream could have an audible effect on the final product. Currently, the most common form of transfer mechanisms installed to meet these requirements are from the professional audio industry, and include AES/EBU and its consumer version S/PDIF covered in Chapter 4.

Performance oriented

Performance orientation accounts for factors such as end-to-end delay and bit rate. For video streams, delay variations in excess of 500 µs are considered annoying with variations in excess of 650 µs being intolerable.

A multimedia-enabled network must deliver a continuous stream of data that arrives at its destination at a fixed rate, even if the network becomes heavily loaded with multiple users and other data streams.

This category requires the most rigorous QoS, since a low latency time is required, as well as high quality audio. The environment in which the transfer is used might include live concerts, where performance audio is generated at the stage, from where it is transferred to the mixing platform and to the D/A process performed at the amplification stage. From there, the final analogue signal is sent out through the loud speakers or to the monitor speakers located near the performers.

2.6.7 Clarifying QoS

Clarification is now needed, for the last few pages have talked almost exclusively about the term 'quality of service' as if it were a product of networks and audio alone.

From the perspective of the client, the quality of any IT services can be thought of in fairly simple terms. Even though a network is made up of a number of complex parts, each of which may be supplied by a different vendor, the client sees the ability to transfer audio as a single service, not a collection of independent services. The consumer is not concerned with how the service is provided, or what components make the service possible. This is true even when looking at a smaller system, such as the simple transfer of audio between one device and another over a fibre, such as in the common case of making a recording to a digital audio tape (DAT) machine. If the transfer does not work, then the chances are that a component is failing, or is not compatible with other components in the chain. No matter which, the service of audio transfer is failing.

The concerns of the client can be classified into broad headings and questions relating to any installation should be aimed at determining these expectations for:

> Availability
> Performance
> Accuracy
> Cost

Availability

Availability focuses upon the question of whether the client can use the services when the client wishes to use them. This depends upon when the service is required. In the example of the administrative offices of a studio complex, this may be from 9 am until 5 pm from Monday to Friday. On the other hand, the studio rooms themselves may be in operation from 10 am until midnight except at weekends when a 24-hour service is required.

For a video-on-demand service provider, a number of choices may be presented. The provider may choose to offer each movie at specific times only, say, twice a night on weekdays, or five times at the weekend. In the end, the service provider permutates the provision of the service to manageable times, whilst considering the consumers' demands, in order to offer a successful service. Most IT managers are aware of the scheduled hours of operation for the service they are operating, and upgrades and other planned maintenance will be scheduled for other times.

Performance

For IT personnel, performance of a network will generally be measured in terms of packets per second, transactions, response times, or other actual measurements that are available from the equipment within the service. Understandably, the customer is not concerned with such jargon, and describes things in rather less specific language such as 'Does it function at an acceptable speed?' It is the job of the designer or IT manager to turn broad descriptions of requirement into measurements and evidence that the service is performing as expected. The question of speed may be the question of response time in an online transaction system. In another case it might be the time that it takes to move a copy of a file from one office to another, or the time required to load an application to a desktop system from a server.

In an audio system, performance stands out in any definition of QoS as one of the main problems because of the unaccustomed quantity of data and the time dependency that audio intrinsically requires.

Accuracy

As with performance, accuracy is also a significant issue for the transfer of audio. It is enough to say that the demands of the different QoS definitions result in different requirements for accuracy. For instance, nothing less than 100% accuracy will do for the transfer of audio in performance orientation, whilst some compromises may be negotiated when defined as a consumer broadcast for special interest groups over the Internet as described earlier.

A simple example of accuracy would be whether e-mail is delivered to the correct recipient or not. Similarly, in the case of applying transactions to a database, it is essential that the change be applied to the proper version. It can be seen that if services do not accurately perform their functions, high availability and high performance are worthless.

Cost

Although performance and accuracy are directly related to QoS, the cost of the service cannot be ignored. A well-known saying in Information Technology is 'Fast, cheap, good – you can have two out of three' and this can be adapted to the delivery of almost any service.

It is common practice for network vendors to present a number of solutions to an invitation to tender (ITT) for the supply of a service and let the customer choose the most suitable. Three separate proposals are common, with the best offering being the most expensive, but able to offer all of the functionality and speed that the customer has asked for, perhaps with a few extra bells and whistles that the vendor hopes will spark the imagination of the customer. The second will typically match the customer's requirements more precisely and will be the middle offering in terms of cost. The third offering will usually be the bare bones system that barely fulfils the basic customer demands, and will be the one with the lowest cost.

2.6.8 Managing expectations

It is possible with most services to provide whatever it is the customer asks for. However, in order to do so may mean an unreasonable cost.

To illustrate this, imagine a ruler marked out in centimetres starting from 0 as shown in Figure 2.12. From one end to the other

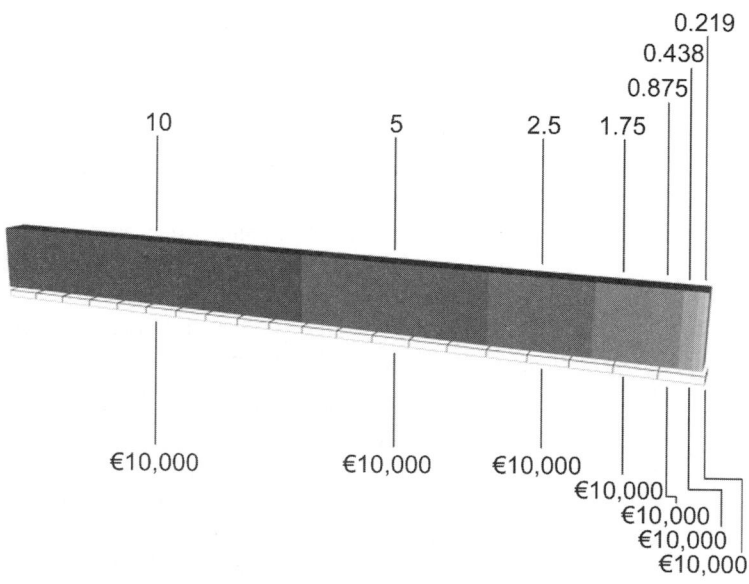

Figure 2.12 Diminishing returns.

end is a fixed value of say 20 cm. A mark is placed halfway down the ruler, marking out two equal lengths of 10 cm. For the next stage, another mark is added exactly halfway between the first mark and 0. Then another mark is made exactly halfway between the second mark and 0. This exercise can be continued indefinitely, with a mark being placed halfway between the last mark and the end of the ruler each time. Although the measurement between the marks decreases by 50% each time, each consecutive mark will be exactly halfway between the last mark and zero. If there were a cost or effort associated with making each mark, then the cost for the project would increase by the same amount for each mark that is made, for increasingly diminished returns, and the objective would never be reached. Using this analogy then, a perfect service that is always available with excellent performance and 100% accuracy is simply not possible.

Notes and further reading

Entry for OSI Seven Layer Model (1999) *Connected: An Internet Encyclopeadia*. http://www.freesoft.org/CIE/index.htm.

Kristula, Dave (1997) *The History of the Internet*. http://www.davesite.com/webstation/net-history.shtml. Various other sources.

Lucent Technologies, Murray Hill, New Jersey, USA (2000) http://www.lucent.com/press/0398/980326.bla.html. Various other sources.

www.whatis.com. Entry by Harbeck, Reg (1999) *What is a Network?* WhatIs.com, 55 West Chestnut Street, Kingston, NY 12401, USA.

3 Practical solutions

3.1 Introduction

In the previous chapter, some of the accepted theories of computer networking were presented along with the historical context in which they were developed. In this chapter, practical examples of common network solutions are explained in relation to theories presented in Chapter 2.

Computer networking covers a wide range of topics from cable types, network topologies, and access mechanisms to protocols, and resilient design and management practices. This chapter presents and explains the most common technologies and mechanisms used to transfer digital data in computer networks. These are related to the 7-layer model where necessary and compared to the QoS that the delivery of digital audio data requires.

3.2 Topologies and cable types

This section explains some of the common technologies and strategies in use for the purposes of data communication. The information given here relates to the physical layer of the 7-layer model.

3.2.1 Topology

In LAN terms, the word topology often causes some confusion, since the logical topology often does not match the physical layout of the cables. For instance, token ring is a ring-based

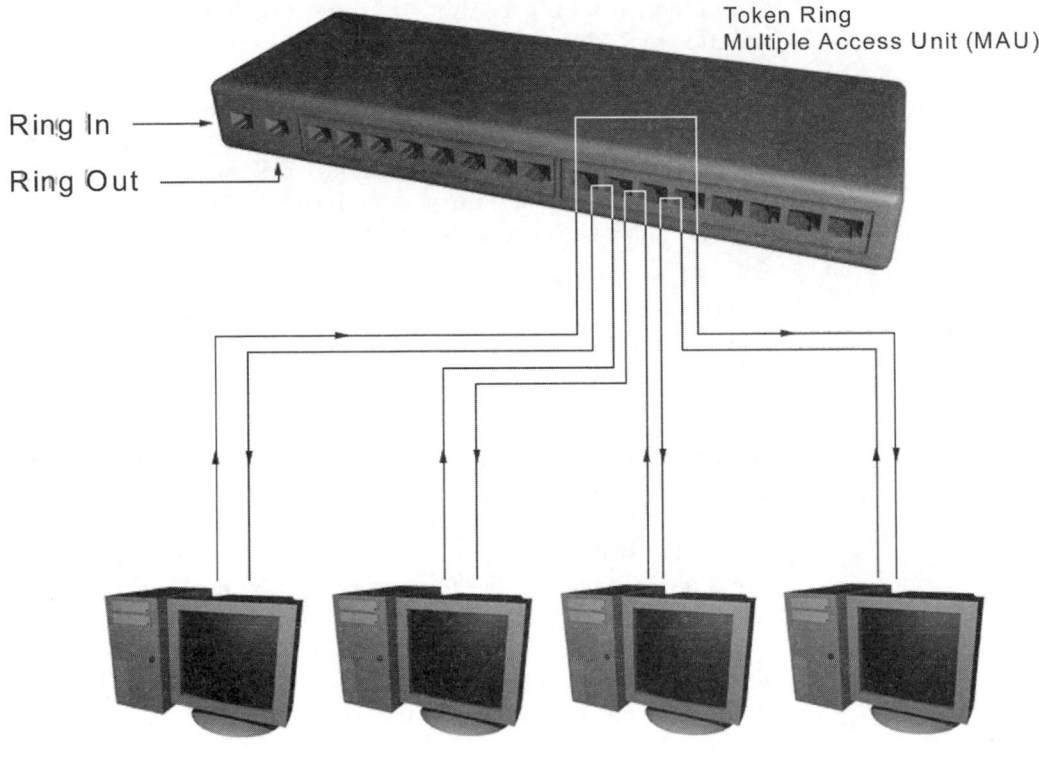

Token Ring
Multiple Access Unit (MAU)

Ring In

Ring Out

Workstations

Figure 3.1 How a ring becomes a star. A token ring multiple access unit (MAU) acts as a hub to centralize connection to the network. Ring in and ring out sockets allow connection to further MAUs. When no other MAU is attached, the completion of the ring is implicit within the single unit. Ethernet (and other) hubs acts in the same way, by centralizing connections.

mechanism, meaning that the cables are connected to form a continuous circle, whilst Ethernet is a bus topology, meaning that the cable is terminated at either end, creating a single network length. However, both systems can be cabled with connecting cables physically extending out from a central hub and so will appear in a physical star formation.

The ring and bus remain, however, and are represented best when drawing a schematic diagram of the connections as shown for token ring in Figure 3.1.

Hubs

A hub is defined as a centre of activity (*WWWebster Dictionary,* 2000) and in the case of network hardware, a hub is where data arrive from one or more destinations and is forwarded in other directions.

Before network hubs became cheaply available, it was common practice in LANs for the main network cable (or backbone) to extend around the working environment, and any new station that needed access to the network would somehow have to access this cable. This has obvious disadvantages with regard to maintenance of the network, as problems occur in work-a-day environments where cables become damaged or broken. Damage to a single network cable can render the entire network unusable.

This problem is compounded because rectifying the problem could involve tracing the entire route of the network from one end to another before finding and replacing the faulty components, a process that could possibly cause even more network failures.

A network design utilizing hubs will usually have a central circuit or backbone, servicing hubs located in one or more locations in convenient positions around the network. This allows the backbone circuit to be reduced to lengths of cable running between hub locations, rather than in physically rigorous working environments where they may be more easily damaged as shown in Figure 3.2. Additionally, by creating a focus for the weaknesses of networks, hubs can be locked away within cabling closets in a controlled environment.

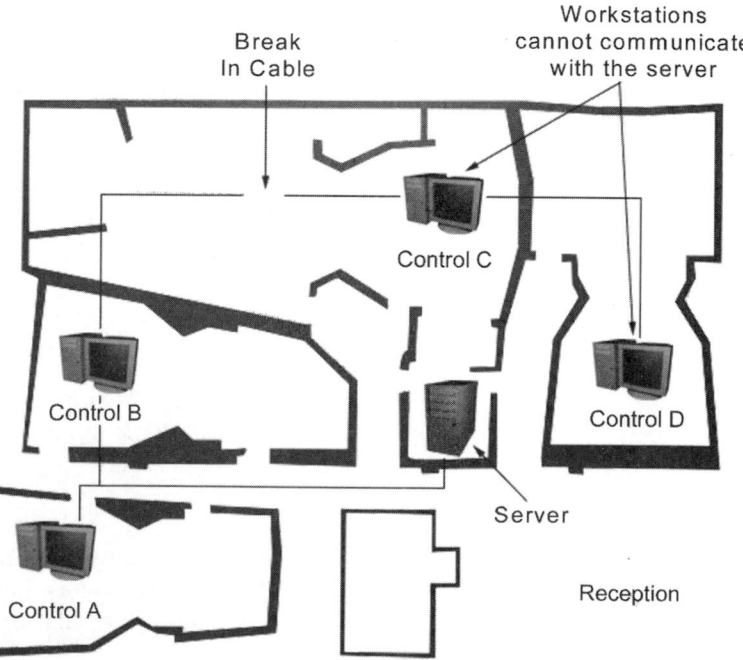

Figure 3.2 This example network layout uses an Ethernet network laid out around the working environment to each device. A break in the cable at any point disables the network beyond and may actually disable the whole network, as the bus is no longer terminated correctly.

Hubs provide convenience as well as the resilience, since damage to a single cable in the client environment will only lead to a problem with that workstation, rather than with large sections (or all) of the network, and faults of this nature are more easily identified and rectified (Figure 3.3).

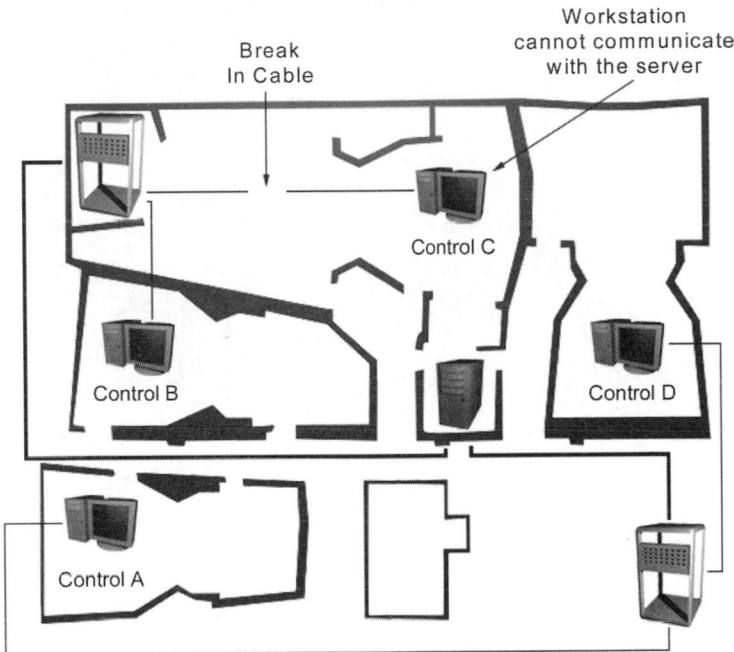

Break
In Cable

Workstation
cannot communicate
with the server

Control C

Control B

Control D

Control A

Figure 3.3 Advantages of hubs: this installation in the same studio has used hubs located in cabling closets to centralize the point of connection. An equivalent break in the cable results in a single workstation being unable to attach to the network.

In slightly larger networks, it is reasonable for a hub topology to consist of a main circuit or backbone to which a number of hubs can be attached (Figure 3.4a). For Internet users connecting through a telephone line, this is the topology used by the ISP, where multiple phone lines connect to a bank of modems connecting to a hub (Figure 3.4b).

Hubs do not effect the functionality of the network in terms of access to data and offer the additional advantage of allowing parts of the network to be reconfigured with minimal disruption. This is particularly useful when used on networks whose access mechanisms do not offer any resilience to disruption, such as Ethernet.

Using traditional backbone cabling principles, client work-stations become detached if any part of the Ethernet segment is disconnected, even momentarily. This is because the attachment of a new workstation would involve disconnecting two parts of

(a)

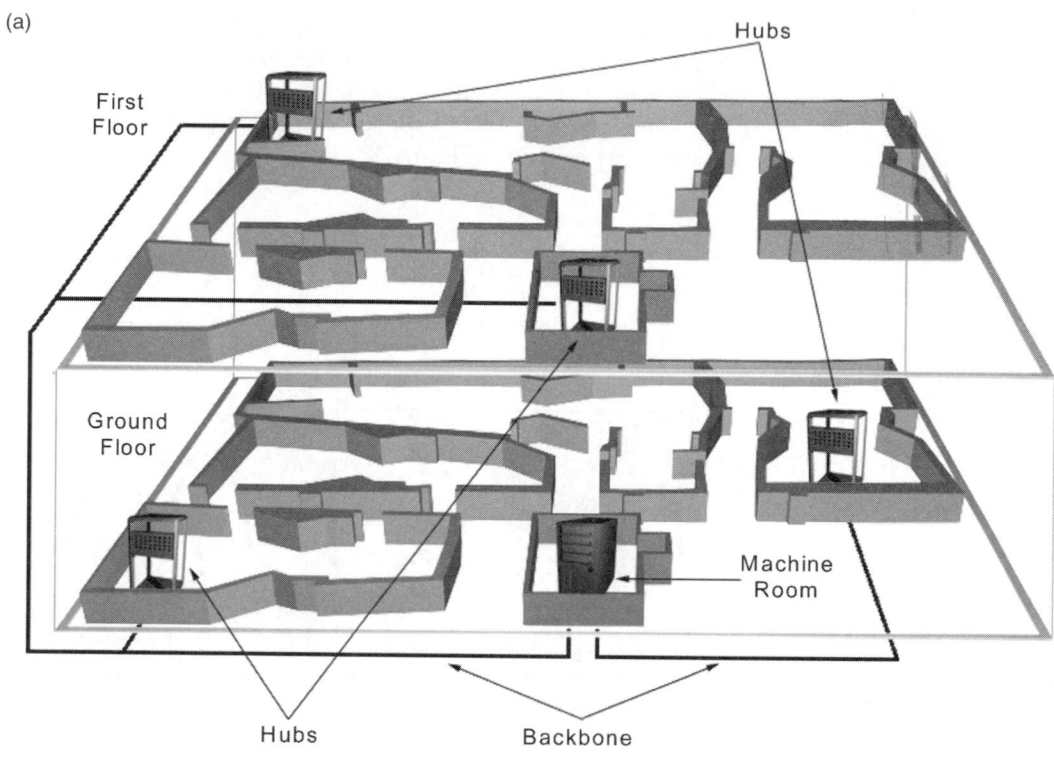

(b)

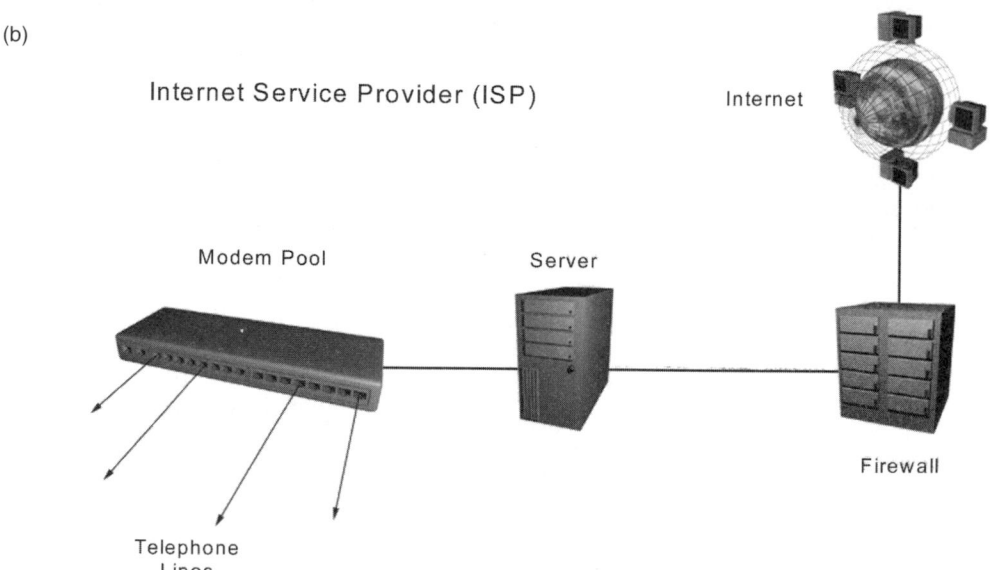

Figure 3.4 (a) Here, the familiar floor layout of the original studio is expanded to a larger facility over two floors. A network backbone connects hubs located in cabling cabinets in convenient locations around the building. (b) Within an Internet service provider (ISP), a hub-like arrangement concentrates inbound connection from the telephone (dial-up) network into a modem pool.

the Ethernet segment and inserting the new workstation into the cabling layout, effectively rerouting the circuit. By disconnecting the segment in this way, one half is no longer attached to the main network, and worse, the remaining half is no longer terminated correctly and also fails.

Hubs avoid this by handling connection to the network in a controlled manner, and isolating those ports that do not have active connections, so allowing attachment to the network whilst it is in operation, or on the fly, without disruption. Instead, reconfiguring the network can be performed through a patch field in a cabinet containing connections to all the points of the network in the same way as a patch field is used to connect equipment together in an audio studio.

Network design annotation

There are different acceptable notations for drawing network designs, as shown in Figure 3.5, which illustrates a network annotation more traditionally, reflecting the logical topology, and

Figure 3.5 Different annotations. This network design is typical of more complex annotations and serves to illustrate the schematic representations typical of such installation projects.

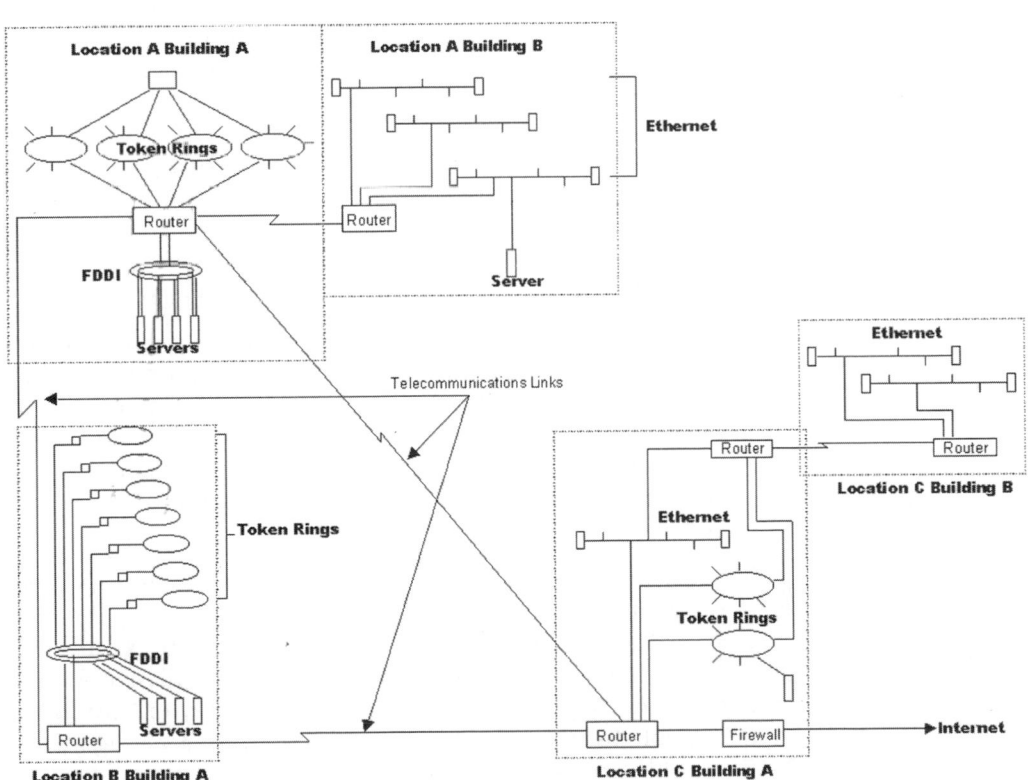

also as a schematic view reflecting the connections between network devices. Other types of network representation are available depending upon the view of the network that is required. These include mapping of the protocol connections, made entirely of logical configurations within the network components, or a mapping more representative of the physical layout of the building into which the network is installed, which might illustrate weaknesses in a design in other ways.

Figure 3.4a shows how the cable layout might look in a larger installation, where several cabling cabinets or closets are dotted around and each house connections to the floor points in their relative area. Each closet will also be connected to the others or to the larger network by using interconnecting backbone cables, also known as vertical cable runs. Vertical is used in this context, since the cabling to the floor points is considered as horizontal and is attached to the vertical backbone, to transport the data down to the equipment room.

It is possible to start imagining the larger picture of the network design process using this information. Each of the inter-connecting cables may need to handle the traffic from several hubs in several closets, each connected to a full complement of floor points positioned to service user devices such as work-stations, printers or other network enabled devices. This means that the interconnecting backbones should be capable of support-ing the accumulated traffic of the user-supporting connections, if bottlenecks are to be avoided.

3.2.2 Cable installation strategies

Significant growth in an organization or department is just one example of why a network might need to be modified. With a simple, non-centralized cabling system, such as might be in use in older or smaller networks as described above, the addition of a workstation to the network requires the installation of a new horizontal cable drop from the workstation to the network backbone.

Most organizations of a reasonable size have a range of systems that need to be interconnected, and should a department need to convert to another system this may mean the removal of the existing cabling and connectors and the installation of new cables. However, for convenience, to save money or because it is impossible to identify redundant cables within a ceiling or floor void, old cable is often not removed, but is cut back and left, resulting in crowded risers, cable trays and ducting.

The time and cost implications of such recabling work are significant. In order to minimize disruption during working hours, major changes are often carried out during evenings, weekends, or scheduled downtimes, adding to the cost of the operation. The solution is to install a standard cable and connector type throughout a building or organization in one large cabling operation. This is known as generic cabling and for additional flexibility, this generic cable is installed and ready for use at all possible locations within a site.

Saturated cabling

Populating an area with generic cabling in this way is known as saturated cabling or flood wiring and is shown in Figure 3.6. The normal guideline is to provide two connections for every 2 or 3 square metres of office space, although requirements vary, depending upon the number of devices that the space is expected to accommodate. Each of the floor points represents one end of a cable that can be connected to the network. The other end of the cable appears in the patch bay, located in the cabling closet that houses the hubs. The exact location is usually determined by a number of factors such as the limitations of the network architecture, the availability of dry vertical risers for connection to the rest of the network and even the availability of enough space to house a closet.

Patch panels are designed to facilitate the easy rearrangement of cable interconnections by centralizing the point of termination. As such, network patch panels are identical in function to those

Figure 3.6 Flood wiring. In this example, the studio has been flood wired, with hubs attached to each other by the backbone. Devices can be conveniently attached to any available floor point. This diagram also shows the vertical cable connection between floors.

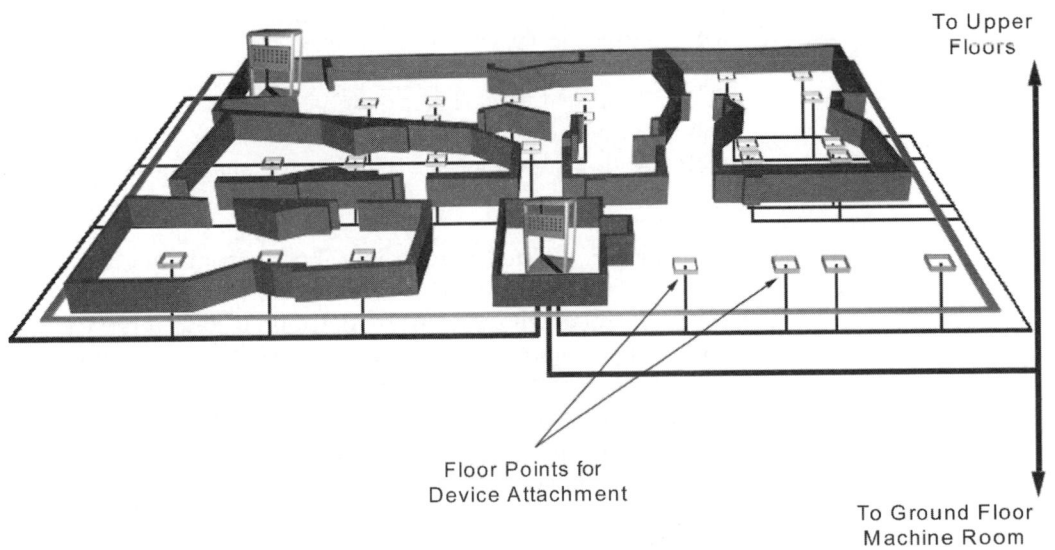

To Upper Floors

Floor Points for Device Attachment

To Ground Floor Machine Room

found in audio studios, which are used for the rerouting of analogue audio signals from one place to another. The network patch panel located in a closet is designed to centralize all the cabling within a particular area (for instance one floor of an office block). These closets will normally contain the network hubs as well as connections to other parts of the network.

The key to a successful structured cable installation is to support the rapid reconfiguration of the network with minimal disruption to the user community. In this way, any of the outlets or wall points can be rerouted to any computer system or service on the network.

The essential characteristics of structured cabling systems are generic cable types, flood wiring, and cross connects. Normally, a standard cable type such as category 5 cable will be selected to fulfil the particular requirements of the client. Cable categories standardize on transmission properties. In larger buildings, cabling closets contain connections between the horizontal and vertical cabling. Once installed, the intention is that no further recabling would be required within the user area. When a new device needs to be attached to the network, the appropriate connection is made between the hub and the patch panel, resulting in no perceivable disruption to other users of the network.

The installation of a generic cabling system implies that the cable type is standardized, allowing all cables to interconnect without the need for media filters or media adapters which are used to translate the electrical requirements between one cable type and another.

Twisted pair cable has been widely accepted for structured cable solutions because of the availability of filters and adapters, as well as the stability and bandwidth that these cable types support.

Category 5 twisted pair cable has become so common in structured cable installations that networking standards, such as Ethernet, token ring and fibre distributed data interface (FDDI) and asynchronous transfer mode (ATM) have been modified to permit its use. The need for structured cabling to support high speed data communication resulted in cabling vendors working to enhance the performance of structured cabling products such as cable, connectors, and cross-connects, to deliver higher speeds, greater distances and improved immunity to electrical interference.

Standards and administration

The Electronics Industries Alliance and the Telecommunications Industry Association (see Notes and further reading) are American National Standards Institute approved standards making bodies. The international standard EIA/TIA 568 from these organizations is described for high speed structured cabling systems recommendations and utilizes a 4 pair unshielded twisted pair (UTP) category 5 cable with connections and data-ports wired to the specifications described in the T568A or T568B documents.

Cable categories 1 to 5 exist, and these are defined by standards from various bodies such as the EIA and the TIA, who produced the TIA/EIA-568-A category 5 standard, amongst others. The categories are classified by the transmission parameters. For instance, category 3 cable specifies cables and connecting hardware with transmission parameters characterized up to 16 MHz, whilst category 4 specifies up to 20 MHz and category 5 up to 100 MHz. The full standard has recommendations for cable systems entering a building, specifications for cabling closets, as well as horizontal and vertical cable runs.

The use of category 5 cable and other components meeting the specifications provides bandwidth capabilities for 100 Mbits/s systems such as the twisted pair implementation of FDDI (TP-PMD), 100baseT and 100baseVG, Fast Ethernet systems and even asynchronous transfer mode (ATM) at 155 Mbits/s.

3.2.3 Cable types

Although category 5 presents itself as the ubiquitous cable type, when installing a generic cable system for the first time, the cable types and their uses need to be considered properly. Several cable standards exist and cable technology improves in the same way as other technologies. The most common cable types are coaxial and twisted pair. For network installations, and the general carrying of digital information, fibre optic cable also needs to be considered.

Coaxial

Coaxial cable, usually shortened to coax, is shown in Figure 3.7 and consists of a copper core embedded in a thick insulator. An electrical shield which acts as the second conductor and is usually made of copper or aluminium braid, surrounds the core. Coax is expensive and is physically quite rigid but is the traditional cable of choice for low cost, small user networks mainly due to its ease of use and reliability.

Figure 3.7 Coaxial cable construction. In this diagram, the copper braid is shown as a mesh, typical of IBM type 1 cable. Phono or RCA type cable, on the other hand, uses simple strands, twisted around the core.

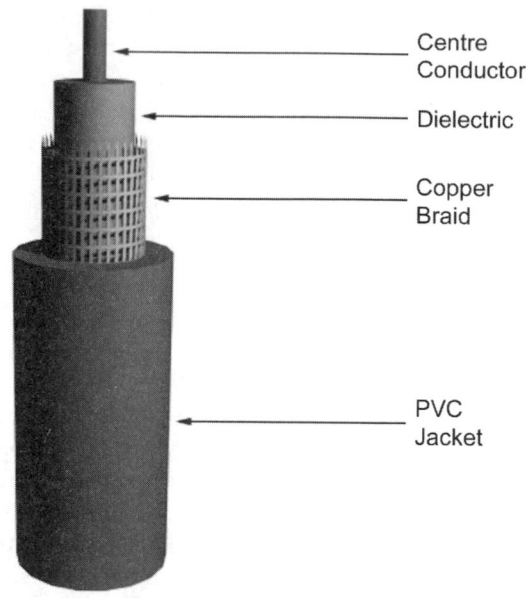

Centre Conductor

Dielectric

Copper Braid

PVC Jacket

The general features of coaxial cable are:

medium capacity
slightly more expensive than UTP
more difficult to terminate
not as subject to interference as UTP
needs care when bending and installing
10Base2 uses RJ-58AU (also called Thinnet)
10Base5 uses RJ-11 (also called Thicknet)

In the early days of computer networks, the popularity of coaxial cable came about because of the original Ethernet cabling types, known as Thinnet or 10base2, and Thicknet or 10base5. The unusual naming convention uses the bandwidth that the cable is capable of supporting, and the length of cable that can support this bandwidth, in hundreds of metres (rounded up to the nearest 100). Using this system, it is easy to recognize that both can support data communications up to 10 Mbits/s over different lengths: 185 metres for Thinnet (10base2, for 10 MB/s over 200 metres) and 500 metres for Thicknet (10base5 for 10 MB/s over 500 metres).

A word of caution is required for the uninitiated, in that the specified maximum length of cable that will successfully transfer digital information is not necessarily the same as the maximum length of cable that the transport mechanism (such as Ethernet or token ring) will sustain. The longest cable run should adhere to the lowest of these figures.

The connectors used in Thinnet Ethernet LANs to join cables together and attach workstations are called T connectors (Figure 3.8) and additionally terminators are placed one at each end of the cable.

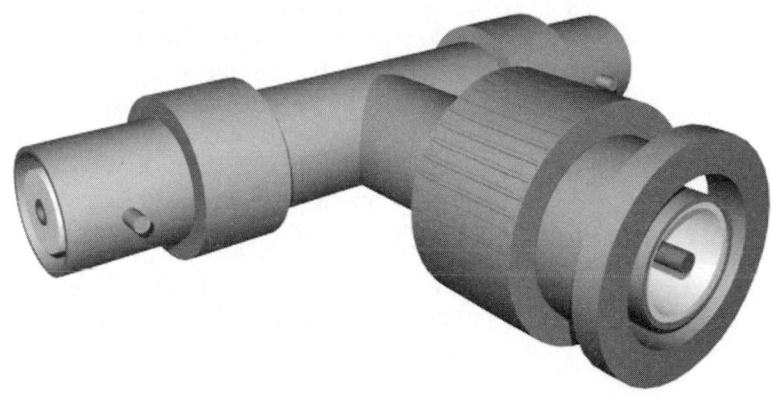

Figure 3.8 Ethernet T piece used for branching early Ethernet networks. Found attached to the network card in the device amongst other positions around the network.

Coax cables should always be joined using the proper connectors. T shaped connectors add branches to the circuit. Every connector in a circuit, however, causes some deterioration in the electrical quality of the signal.

Twinaxial

Twinaxial or twinax cable shares most of its construction with coax cable, except that there are two conductors in the core instead of one. Insulating material independently surrounds each of the two conductors. This cable type was mostly used for connection to proprietary IBM systems, but became more popular as IBM's token ring network access mechanism also became popular in commercial computing environments (Figure 3.9).

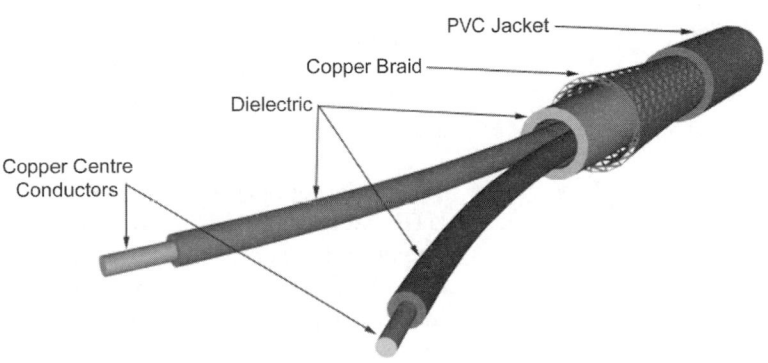

PVC Jacket

Copper Braid

Dielectric

Copper Centre Conductors

Figure 3.9 Twinax cable construction.

Twisted pair

Category 5 cabling is most commonly installed as the cable choice for new medium to large network installations. Category 5 or cat5 cable is an example of twisted pair cable and is particularly used within structured cable installations. There is also a sister standard known as Level V, and although the two types are not identical, both support up to 100 Mbits/s data transmission. The physical cable assembly requirements are the same, and both can be referred to as 10baseT. One of the most common problems which a network engineer may face is the correct way to make up a 10BaseT cable. Usually, just to add to the problems, there is a requirement for a reversed or cross-over cable, which is usually used to connect devices directly.

Twisted pair comes in either shielded or unshielded formats and both consist of four pairs of wires that are manufactured with the wires twisted to certain specifications.

Cat5 cable is an unshielded twisted pair (UTP) cable consisting of eight wires in four pairs. The strands that constitute each wire will either be a single strand or multiple strands, referred to as solid or flex, respectively. Typically the solid cable type is used during the generic cable installation operation and runs through walls and ceilings. Flex, on the other hand, is more flexible and is used in the working environments to make shorter drop cables located between the wall plate (or floor point) and the device itself, and for patch cables used in the cabling closet to connect the patch panel to the hub.

A further consideration during the cable selection process regards the exterior sheath, or jacket, which may be plenum grade or non-plenum grade, referring to their fire codes. This may be important if the cable run traverses any fire partitions within the building.

The pairs of wires in UTP cable are colour coded by pair, so that the pairs can be identified from end to end. Typical cat5 UTP cables contain four pairs made up of one wire of a solid colour and the matching wire in the pair is the same colour striped onto a white background.

The cable connectors and jacks most commonly used with cat5 UTP cables are RJ45, shown in Figure 3.10. RJ stands for Registered Jack and the 45 designation specifies the pin numbering scheme. The jack is the male proportion of the assembly and the female part of the assembly is sometimes referred to as the Jill for ease of reference.

Figure 3.10 RJ45 connectors showing male and female versions. The female version of the connection is generally mounted within a device or patch field in a cabling cabinet.

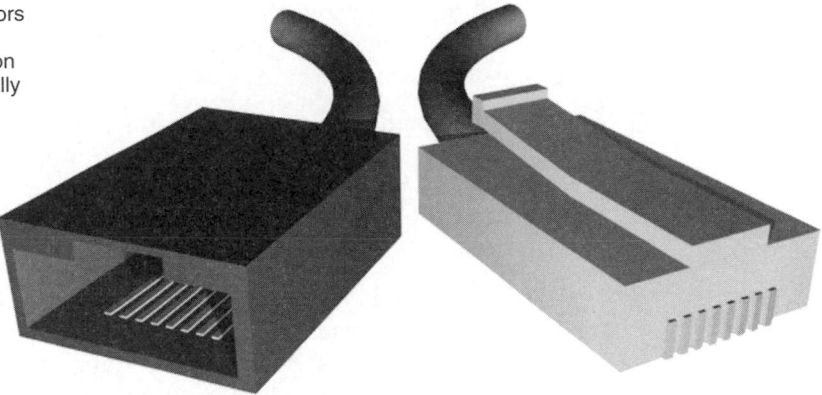

The IEEE Specification for Ethernet 10BaseT requires that two twisted pairs are used and that one pair is connected to pins 1 and 2, with the second pair connected to pins 3 and 6. Pins 4 and 5 are skipped and are connected to one of the remaining twisted pairs. Figure 3.11 shows the pair numbers and pin-outs.

Figure 3.11 Table showing pair numbers and pin-outs.

Pair 1	
Pin 4 wire:	Solid blue
Pin 5 wire:	White and blue stripe
Pair 2	
Pin 1 wire:	White and orange stripe
Pin 2 wire:	Solid orange
Pair 3	
Pin 3 wire:	White and green stripe
Pin 6 wire:	Solid green
Pair 4	
Pin 7 wire	White and brown stripe
Pin 8 wire	Solid brown

Fibre optic

Fibre optic cable is considered the default choice for connections involving high data rates and is particularly prevalent in audio installations where the large quantity of data requires a cable type that is up to the job. Fibre is also useful in installations with long distances and interconnecting networks. Fibre costs more than either twisted pair or coax, and requires special connectors and joining methods.

To terminate a fibre optic cable, the face of the cable must be exactly perpendicular to the cable length in order for the light to exit gracefully. This manufactured termination must also sit exactly flush with the connection to the device, in order that the signal can be received without error.

Fibre optic cable systems tend to be expensive, but the high capacity bandwidth makes it particularly useful for the network backbone.

Since many horizontal segments may be attached to the vertical backbone, the data rate for the components that make up the vertical run (including the transport mechanism) should be equal to or greater than the sum of all the horizontal networks attached to it, if bottlenecks are to be avoided.

Fibre should also be considered in areas where the amount of RF is either extremely high or needs to be kept extremely low. Since fibre uses light to transport data, it is unaffected by such interference and does not generate any. This is useful in controlled environments and in areas such as in aircraft or specialized RF anechoic chambers used for testing emissions, where fibre is the preferred cable type.

Fibre optic is also used to overcome distance limitations since good quality fibre is capable of transmitting data over several kilometres without degradation in the signal and so can be used to join networks between buildings.

Fibre optic cable uses differences in light intensity rather than any electrical characteristic to represent binary digits. The cable itself consists of a protective outer sheath and cladding, enclosing the glass or plastic transmission core as shown in Figure 3.12.

Plastic core fibre optic cable is less expensive than glass, but is also subject to more loss of signal than glass, due to the inconsistency of the molecular structure of the conductor, and it therefore cannot transmit data over as great a distance.

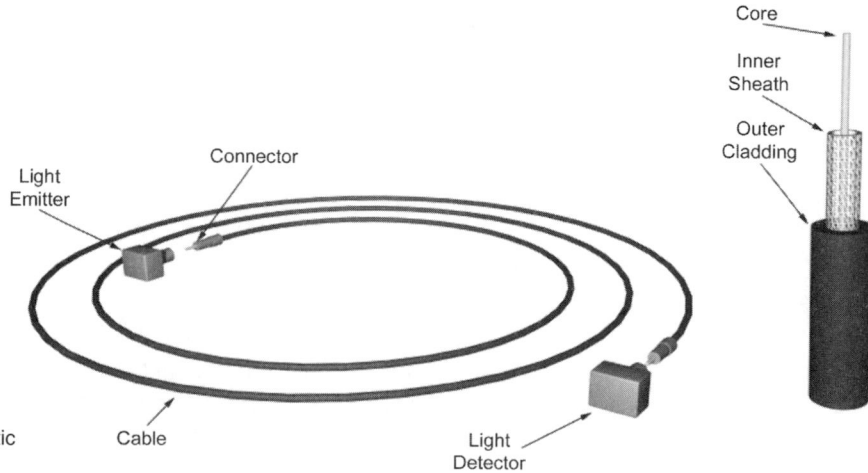

Figure 3.12 Fibre optic cable construction.

Bend rating

The bend rating (sometimes known as the bend ratio) is a number applied to fibre optic cables and some other cable types such as coax. The number represents the distance in metres in which the core or multi-core cable can perform a 90 degree turn without fracturing or stressing the conductor. Fibre optic cables with more cores have a higher bend ratio, meaning that the space required for a 90 degree bend is higher. Plastic core cables have a smaller bend ratio than glass.

Using just distance as a measure, glass core is preferable although glass core has a higher bend rating and is subject to problems of impact and shock, where the glass may become damaged inside the sheath, resulting in poor or broken transmission.

Slight faults in fibre cable are notoriously difficult to identify and rectify due to the nature of the installation. As a result, network designers can specify that spare fibre cores are installed during the installation so switching cores within the multi-core cable run can easily rectify faults. Increasing the number of cores in a cable run in this way increases the bend rating of the cable.

The bend rating may render fibre optic cable as unusable in some locations, especially those with limited space.

Fibre tends to be associated with FDDI and Fibre Channel network access mechanisms, amongst others.

3.2.4 Other media

Transport media is not limited to physical cabling, and it is common for commercial organizations to utilize other media for the transport of data. Several techniques are available depending upon the circumstances.

Within a small enclosed area such as an office or studio, several manufacturers offer infrared connections to the network. These take the form of ceiling mounted devices, which send and receive data from infrared transceivers connected to (or as part of) the media access card within each device within range. The ceiling device is connected back to the cabling closet, where it connects to the rest of the network (as shown in Figure 3.13). The ceiling device acts as a hub, with the advantage that no additional patching is required whenever a new device is attached to the network. The early limitations of this system were that each device must have a line-of-sight to the ceiling hub, meaning that no solid objects must be placed between the transmitter attached to the back of the workstation and the ceiling hub.

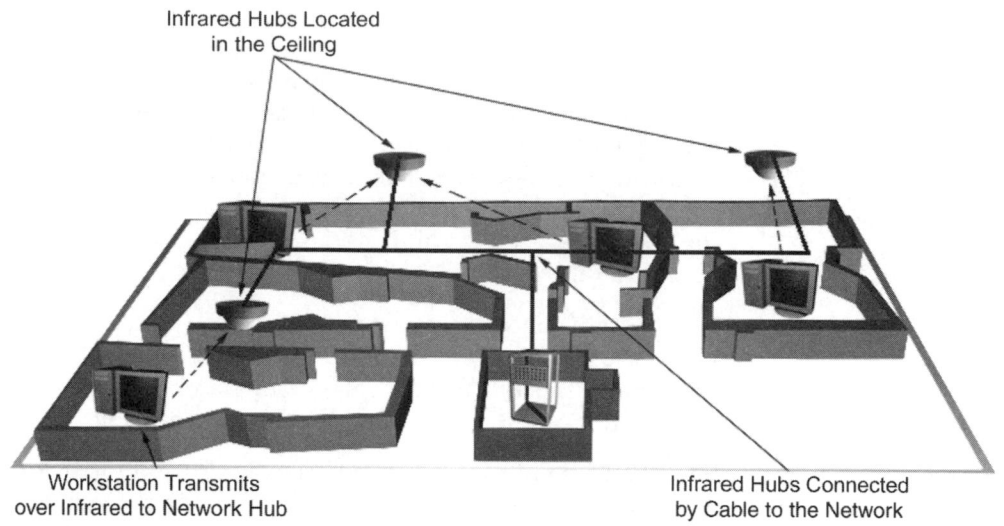

Infrared Hubs Located
in the Ceiling

Workstation Transmits
over Infrared to Network Hub

Infrared Hubs Connected
by Cable to the Network

Figure 3.13 Using infrared
Ethernet controllers, cabling
to each device is discarded
and replaced with cabling to
infrared hubs, one located in
each room or enclosure.
These service devices are
located within the proximity,
although it should be noted
that a line-of-sight may be
required between the device
and the infrared hub.

Over longer distances, other techniques exist, which include line-of-sight RF transmission, particularly suitable for making network connections within organizations with a number of sites geographically displaced within a municipal area.

Satellite transmission can also be counted in this section, for global area networks and installations requiring high data rates. Satellite communication is intrinsically expensive to operate and unless the owner of the network is prepared to buy a satellite, some or all of the service will be managed in some way. Satellite communications are covered in more detail in the Chapter 8, relating to telecommunications networks.

3.3 Network access mechanisms

This section deals with the most common methods for accessing networks and relates to the media access and logical link sub-layers within the data link layer of the 7-layer model.

3.3.1 Dependency on adjoining layers

The cable selection process not only depends upon the design of the building, the placement of network facilities, and environmental considerations, but also upon the network access mechanisms that will be used.

The QoS is dependent upon the whole network service from the amount of time that the network is not available (known as a network down) to the data capacity and reliability of the bit transmission mechanism. Therefore, installing a cable type capable of supporting 100 MB/s when the network access mechanism cannot deliver the QoS requirements will be met with some objection by the person or organization responsible for paying!

3.3.2 Ethernet

Ethernet is the most widely used network access mechanism according to sales of network access cards, although the tendency is to group the various types of Ethernet together in the same category.

History

The original Ethernet specifications came from a commercial initiative between three major technology vendors, Digital, Intel, and Xerox. The initiative became known as DIXI and produced the Ethernet Blue Book specifications at around the same time as the DoD 4-layer model was produced. The Blue Book describes a method of accessing the cable, called carrier sensing multiple access with collision detection (or CSMA/CD). This was finally published between 1979 and 1980 from the famous Xerox sponsored Palo Alto Research Center (PARC). As mentioned earlier, this was not the only groundbreaking idea to emerge from this establishment, as the first GUIs were also attempted here.

Standards and administration

The CSMA/CD concept was made available for general consumption, an easy-to-understand technology for high speed data communications between digital devices, and became the basis for the Ethernet standards (RFC894), IEEE 802 (RFC1042) (The Institute of Electrical and Electronics Engineers – see Notes and further reading), and ARCNET (RFC1051), all of which are well-known groups of networking standards.

General description

In simplified terms, the Ethernet access mechanism works by using a single pair of wires to connect each device on the network. Each device requiring attachment to the network must connect directly to this cable. Each device on the cable has a unique address known as the media access control (MAC) address.

To understand how Ethernet works, imagine a cable length with three computers attached to it, as shown in Figure 3.14. Computer A sends a message to another computer, by broadcasting the message throughout the entire length of the cable.

Figure 3.14 (a) How Ethernet works. Devices wishing to transmit messages do so throughout the entire length of the cable. (b) When two devices attempt to send messages at the same time, the messages collide. As a result, one or both devices send a jamming signal. Problems can occur if the cable is too long, resulting in the device failing to determine that a message being sent has collided with another message.

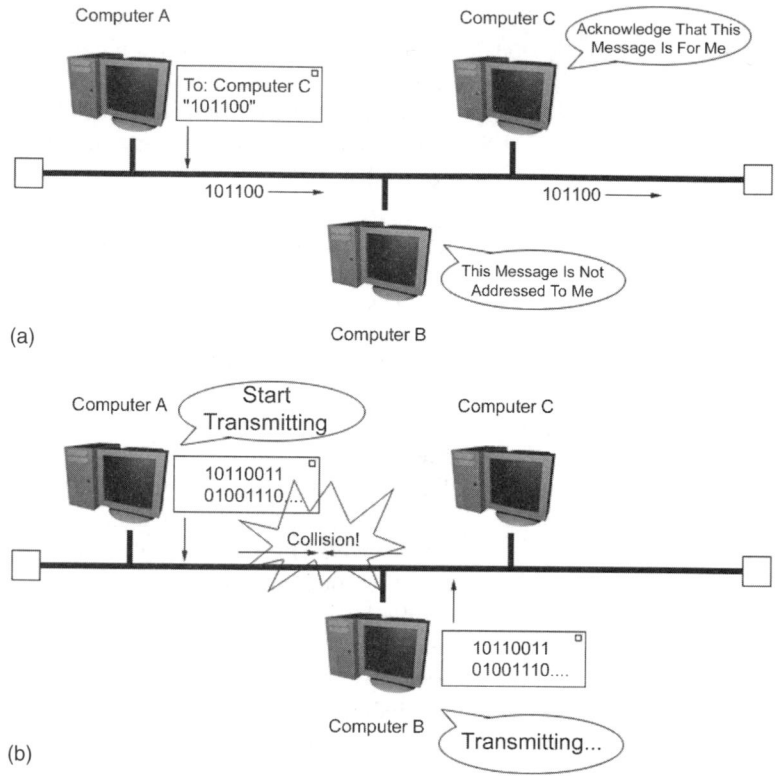

Each device listens to the cable bus constantly and receives the message. When a message is broadcast, each receiving device compares the address at the start of the message to its own address, in order to ascertain whether it is the intended recipient. If it is not the intended recipient then the device will ignore the message. If the message is intended for that device, then the message is copied into the upper layers for further action.

Problems occur when two devices want to transmit data at the same time, and this is where collision detection is invoked, and this is illustrated in Figure 3.14b.

Supposing that computer A wants to send out a message on the network but computer B is already in the middle of sending a message. Computer A will hear the message and wait until computer B has completed its transmission before attempting to send its own message. In Figure 3.14b, computer C is also waiting to send a message, and does so as soon as computer B has finished. Since A and C are located physically distant from one another, it takes a fractional moment for the message to arrive at the other device. The result is that both devices start to transmit when computer A has finished. As a result, the messages from both computers become garbled by the collision between the two transmissions. Both devices continue to listen to the cable during transmission and when a collision is detected, a jamming signal is generated by one or both of the devices. The jamming signal is sent before the end of the message has been transmitted, so both computers recognize that the jamming signal has occurred as a result of their transmission colliding with that of another device.

In response to the jamming signal both A and C stops transmitting for an undetermined period of time, before attempting to send the messages again. Each computer then attempts to send their information again, first checking to see whether the cable is free. This process will continue until both stations have succeeded.

The undetermined waiting time is set by a pseudo random number generator within the media access card, and for each successive collision, the random time period is generated from an increasing number.

Strictly speaking, this access method does not require the definition of the cable type and other physical layer parameters and sits happily within the MAC sub-layer of the data link layer.

However, it is important to realize that a collision between two sending stations must be sensed before the end of the message transmission in order for both transmitting devices to realize that they are the affected parties. This means that the cable length between two stations must not be so long that computer A receives a jamming signal as indication of a collision, after it has finished sending its message. If this happens, then the collision sensing algorithm in computer A will assume that this jamming signal pertains to some other collision and would not resend its garbled message. The result is that this MAC sub-layer process defines the maximum cable length, which is normally associated with the physical layer. Since Ethernet describes basic cable lengths, it also details resistance, voltage, and encoding methods, all of which sit in the physical layer.

As one of the first commercial (and ratified) standards for computer communication, we can see that Ethernet follows the argument regarding the 7-layer model, and does not fit strictly into the layered functionality of the model.

Considerations

Looking at established mechanisms such as Ethernet, in terms of the OSI model, creates a better understanding of how the technologies fit into the scheme of things, but does not directly address the issues for the transfer of multimedia, and particularly streamed audio information.

When a collision is detected both stations back off for an undetermined time period. This undetermined part of the mechanism means that Ethernet cannot provide a deterministic QoS as required for some real-time applications. If there were only two stations on the network, and one was streaming audio information to the other, then there would be little problem provided the amount of data being sent did not exceed the available data rate.

The problem does not increase especially if more than one receiving station is introduced, with only one station transmitting, since the number of collisions between sending stations would still be zero. (This is not entirely true when using most common upper layer protocols, as discussed in the section on protocols.) As soon as additional sending stations are added, data packets may start to collide and the streamed signal is at risk of sounding interrupted and of reduced quality.

Network professionals will often claim that a 10 MB Ethernet network segment will only achieve a maximum throughput of 3 MB, depending upon the number of stations, because collisions are taking up so much bandwidth, although the performance graph actually decreases exponentially. This is because Ethernet segments have less and less time available on the network for successfully sending data, and the likelihood of colliding with another station increases exponentially, until the performance ceiling is reached.

It is possible to make Ethernet more efficient and deterministic, but in order to do this the mechanism needs to be arbitrated, to ensure that no collisions are experienced. The new arbitration rule would have to be implemented in such a way as to not interfere with the Ethernet standards, and to ensure that collisions did not occur. The mechanism would have to be implemented throughout the network, since any station without the new rule would adhere to the old rules, and send data as soon as the cable appears to be free. This is discussed more thoroughly in the section regarding CobraNet, which uses a proprietary 'O-persistent' arbitration layer to ensure that no collisions take place. This layer is inserted before the sending mechanism is invoked. In this way, the branch of the Ethernet algorithm that invokes the random wait time is never invoked. Unfortunately, this precludes the use of any other protocol on the network, and so remains proprietary to the network.

3.3.3 Token ring

Token ring topologies continue the trend of sharing the transmission medium but differ from Ethernet in the critical area of the access method by regulating access to the cable.

History and standards

IBM originally developed the token ring network in the 1970s and it remains second only to Ethernet/IEEE 802.3 in general LAN popularity, based on the sales of MAC cards. The related IEEE 802.5 specification is almost identical to and completely compatible with IBM's token ring network modelled, as it was, on the IBM token ring work. The differences are illustrated in Figure 3.15 and it continues to shadow IBM's token ring development. The term 'token ring' is generally used to refer to both IBM's token ring network and IEEE 802.5 networks.

	i	2
a a a	4 or 16 bit s s	4 or 16 bit s s
a i	26 S P 72 P	25
	Star	ot S pe i ied
ia	wisted pair	ot S pe i ied
i a i	aseband	aseband
	o en P assing	o en P assing
i	i erential an hester	i erential an hester

Figure 3.15 Comparison chart for IEEE 802.5 and token ring.

General description

A token ring network uses a special frame called a token that rotates around the ring when no stations are actively sending information. A network of only one device sets the network in single station mode and that device becomes responsible for generating the token. As other devices become active on the network, the first station remains the master, responsible for regenerating lost tokens.

Token ring communicates over four wires: two for transmitting and two for receiving and this can be STP or UTP.

Topology

Token rings are usually installed around a multiple access unit (MAU), which is a complete network ring encased in a passive box. Its appearance and resulting cable layout make it exactly comparable to a hub. Each connection to the MAU includes a relay that is operated when a station is activated, thus opening the connection to become part of the ring.

The robustness of the arbitration and general signalling mechanism means that any temporary disruption to the ring will be quickly recovered.

Two further connections are always supplied on MAUs, these being ring in (RI) and ring out (RO). These can be connected to their counterparts on further MAUs in order to extend the ring.

A ring is constructed by connecting the RO of one MAU to the RI of the next MAU in the ring, as shown in Figure 3.16. The last MAU in the ring connects back to the RI of the first MAU.

A ring can consist of a single MAU, and in this case there is no need to connect the RI and RO as the connection is implicit.

Arbitration

Instead of allowing all devices to transmit at once, token ring requires a process of arbitration before transmission can begin. This orders the access process and provides a deterministic QoS.

When a station wants to transmit on the ring, it must first capture the token, thus becoming the owner. Once a device is the owner of the token, it can transmit.

In order to transmit, the device appends data to the token, changing it from a token into a frame. As the frame traverses the ring, it passes through each station in turn, on the way to its destination.

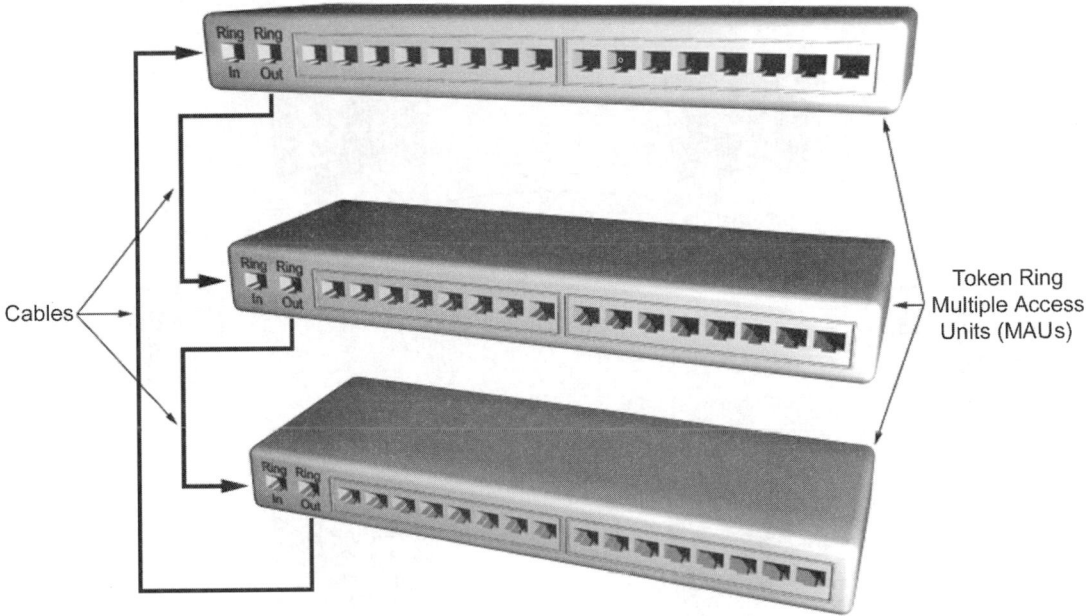

Cables

Token Ring
Multiple Access
Units (MAUs)

Figure 3.16 Construction of a token ring showing the cable connections in a three MAU ring.

Error checking

Each station receives the frame and inspects the data to see whether it is the intended recipient. If not, then the frame is regenerated and placed back onto the ring. An error check occurs each time a station repeats the frame. If an error is found, a special bit in the frame called the error detection bit is set so that other stations will not report the same error. Once the data arrive at the destination station, the frame is copied to the destination's token ring card buffer memory.

Acknowledgement

The destination station repeats the frame onto the ring, changing two series of bits on the frame. These bits, called the address recognized indicator (ARI) and the frame copied indicator (FCI), indicate that the destination station has received the frame without error. The frame continues around the ring, arriving back at the source station. The source station recognizes the sending address as its own, and strips the frame from the ring. The payload is inspected and compared against the version in memory as a final integrity check.

Provided the process has occurred without error, the source station drops a free token onto the ring and the process starts again. The entire process is illustrated in Figure 3.17. If an error has occurred, the data are requested again.

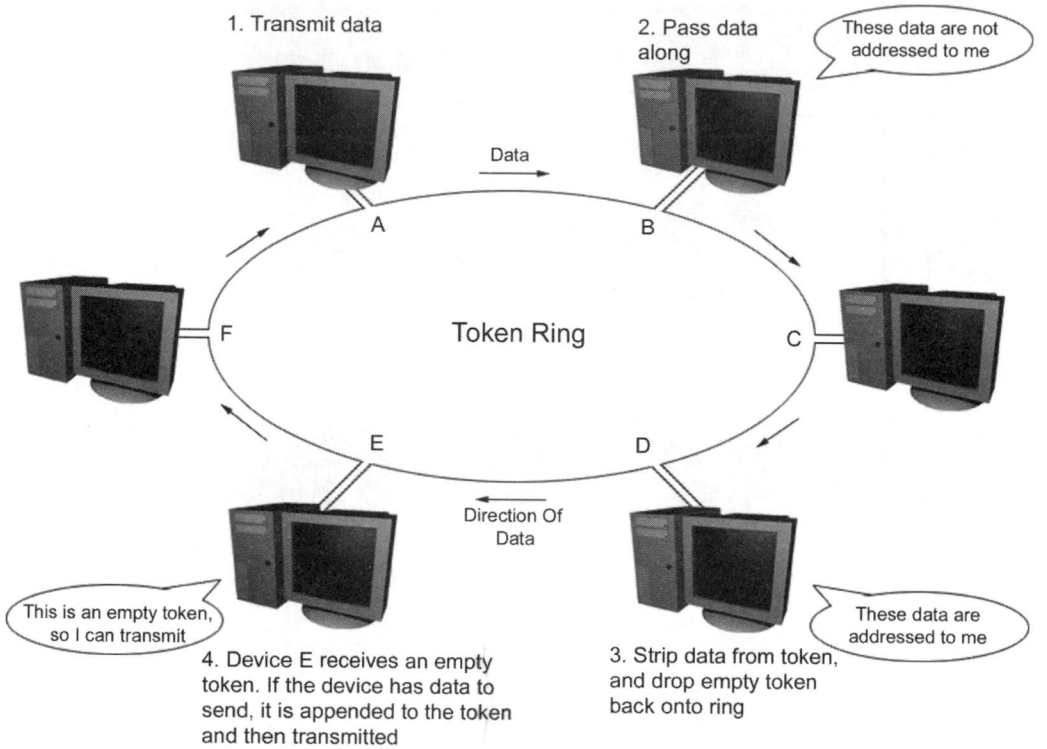

Figure 3.17 Token ring signalling. Device A transmits successfully in Stage 1. Any interceding devices (such as devices B and C) recognize that the data are not intended for receipt and retransmit the full frame token.

3.3.4 Fibre distributed data interchange

Fibre distributed data interchange (FDDI) is an early technology capable of offering higher data rates to general user community workstations for a reasonable amount of money.

FDDI is a ring-based token-passing technology, and most commonly uses fibre optic cables running at 100 Mbits/s, although copper versions are now available, running at the same speeds.

History

FDDI comes under the same administrative umbrella as Fibre Channel, covered in Chapter 6, although the first draft proposal was produced as early as 1982. FDDI followed the conceptual ideas of IEEE P802 Standards Project for LAN and selected the 4 Mbits/s–token ring protocol (P802.5) as a starting point.

Standards and administration

The requirement for a technology such as FDDI was originally conceived within the ANSI approved X3T9.5 task group for the development of standards concerning serial interfaces.

As with so many complex definitions, FDDI is made up from several standards, such as the specification for physical protocol sub-layer ISO9314-1 PHY, which defines the signal encoding and decoding mechanisms, clock synchronization mechanisms, clock rate and signal waveforms for transmission over FDDI media. The media access control sub-layer, defined in ISO93412-2 MAC, defines the protocol for station to station communication, as well as the frame formats and synchronous and asynchronous access.

General description

Apart from the early speed:cost ratio, another attractive aspect of the network access mechanism is its ability to recover from failure. This is achieved by using two separate rings, known as the primary and secondary rings. Network devices requiring attachment to the ring do so through the primary ring, where data transmission takes place.

Devices attaching to the network are classified according to whether they attach only to the primary ring or to both the primary and secondary rings. Stations attaching to both rings are known as dual attach stations (DAS) and those attaching only to the primary ring are known as single attach stations (SAS) (see Figure 3.18).

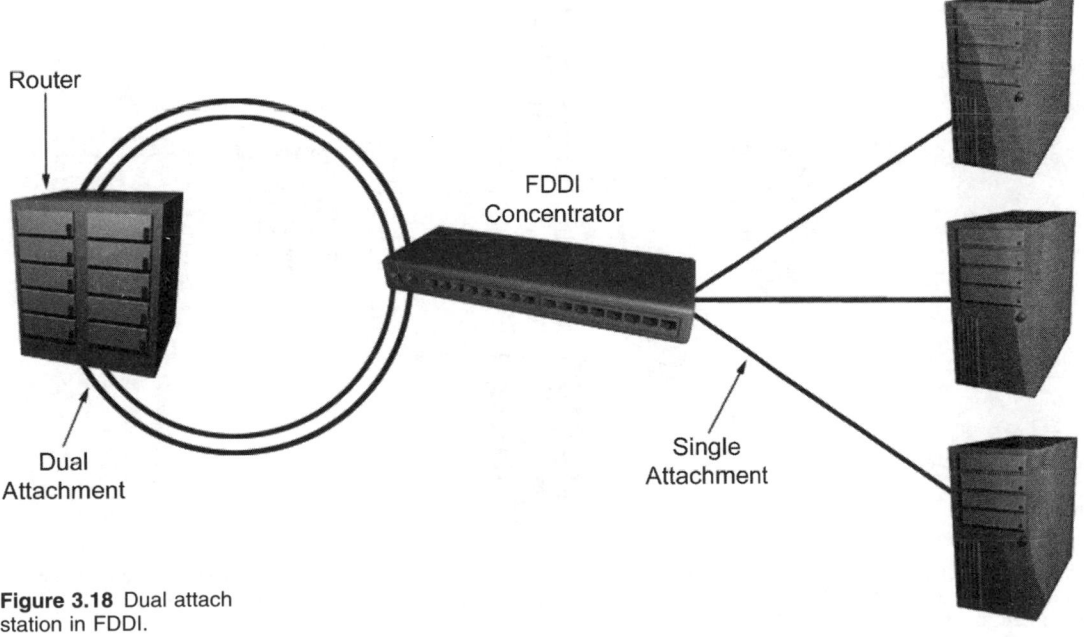

Figure 3.18 Dual attach station in FDDI.

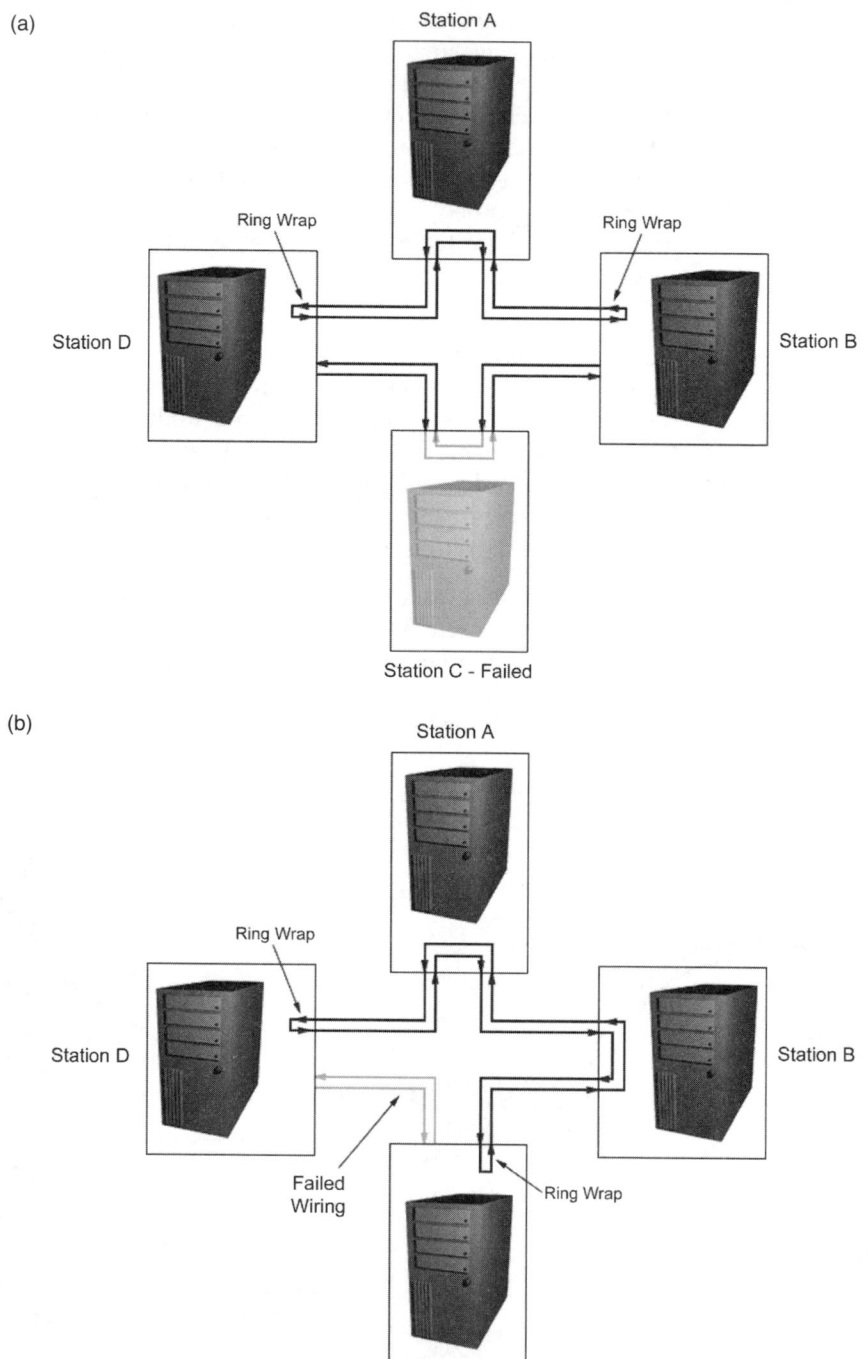

(a)

Station A

Ring Wrap

Ring Wrap

Station D

Station B

Station C - Failed

(b)

Station A

Ring Wrap

Station D

Station B

Failed
Wiring

Ring Wrap

Station C

Figure 3.19 Link failure recovery in FDDI. (a) Station C has failed in this example, and so FDDI wraps the ring at stations D and B. (b) In this example, the cable has failed, rather than the device, and so service can be resumed to all devices on the ring.

The secondary ring acts as a backup system. If a failure should occur in the primary ring, the nearest DAS will wrap onto the secondary ring in order to isolate the fault and maintain connectivity. In this case, the network continues to operate in single ring mode until the fault is rectified as shown in Figure 3.19.

Topology

Like other network topologies, FDDI networks use a centralized unit through which other devices must attach. Essentially performing the same function as MAUs in token ring networks, in FDDI terminology, hubs are known as concentrators. Concentrators are provided in single-attach and dual-attach formats. As with token ring, a network consisting of one concentrator does not require external connections, as the ring will be provided on the backplane of the concentrator.

The ability of FDDI network components to be attached to two separate rings is exploited in the provision of dual homing. Dual homing means that one concentrator is attached to two other completely separate concentrators as shown in Figure 3.20.

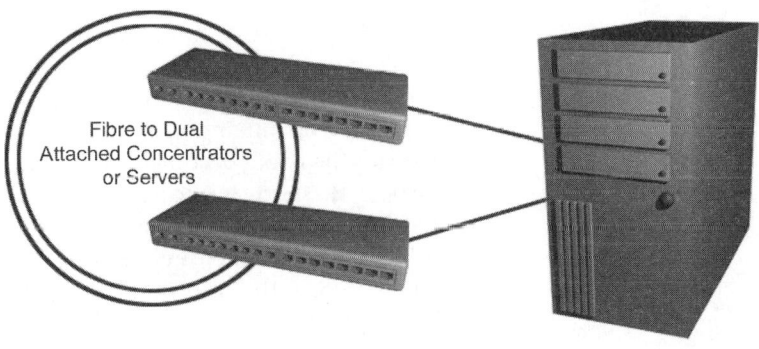

Figure 3.20 Dual homing in FDDI, showing a single server attached to two separate FDDI concentrators. Subtly different from Figure 3.18, in this case the workstations can be attached to two separate hubs.

Fibre to Dual Attached Concentrators or Servers

Media access

The MAC layer is responsible for capturing and retransmitting the token, and two types of token, restricted and unrestricted, are defined.

The unrestricted token is available to all the stations on the network, whereas the restricted token is used when a station requests a prolonged conversation with another device on the network that requires all (or nearly all) of the unallocated bandwidth on the network (*FDDI Basics* – see Notes and further reading).

3.3.5 Considerations for local area network installations

It is a common understanding that the efficiency of Ethernet networks declines as stations are added and traffic increases, since collisions take up an increasing amount of the available bandwidth, and busy 10 Mbits/s Ethernet networks will struggle to deliver more than 3 Mbits/s in these circumstances.

Token ring, on the other hand, has the ability to set a priority for stations that need faster services, although in practice this is usually reserved for host machines or other major network components, which might otherwise create a bottleneck. In addition, token ring offers a more controlled method for accessing the network and provides greater efficiency as more stations are added, but there is also a small management overhead of 6–10% to consider contained within the framing of data.

On the whole though, token passing can provide a deterministic QoS, making it worthy of consideration as a candidate for the delivery of streamed media, although close management and configuration are required.

The ability to create a token bus network is used to effect in ring topologies, such as token ring and FDDI (fibre distributed data interface) networks, where a broken cable will not result in the whole network being unavailable to devices wishing to access it. Instead, the cable becomes a bus topology, allowing at least some parts of the network to remain usable. In this case it is possible that both separated parts of the network remain usable, whilst unlikely that all the devices on each half of the network will be able to access the information required, unless the network was so designed.

The performance of a token passing network is determined by the time it takes the token to go around the ring, plus the time taken to perform the actual transmission, known as the transmission delay.

The number of nodes that the token must pass through determines the size of a token ring. The maximum delay that any one station experiences before it receives the token is calculated using the ring size and the maximum packet size. Consequently, token ring networks can be described as deterministic and can be used in applications with specific QoS requirements for delivery, whereas Ethernet cannot.

3.4 Protocols

The modularity of the 7-layer model serves to illustrate that the type of information being transferred over a network is not necessarily of concern to the mechanisms that move it. However, as with network access mechanisms dealt with by the physical layer and media access sub-layer, some mechanisms are not suitable for transfer of particular types of data. As such, the choice of protocol is just as important to the overall QoS as other layers.

This section discusses implementations at the network and transport layers of the 7-layer model, and generally describes protocols.

3.4.1 Introduction

Protocols wrap around the data itself, and contain information pertinent to the transfer of the data. In all cases, this extra information has to be stripped off before the data can be used, or passed upwards in the model.

Both the 4-layer and 7-layer models describe the steps that are encountered when interconnecting networks.

The methods describe the resilient delivery of data packets and allowed for multiple connections over disparate geography and media in order to deliver data. The functionality implemented at this level allows data to be routed through pathways by decision-making algorithms within devices known as routers. If any link between two routers is found to be unavailable, the mechanisms within the network layer can ensure delivery by another route, if one is available. The functionality required to route data packets around problems in this way is further enhanced so that a different route can be utilized depending on the status of particular pathways at any one time.

For instance, if one route becomes congested for a period of time, there may be other routes around the traffic jam which, although longer, will provide quicker delivery whilst the shortest route remains congested, as shown in Figure 3.21.

In terms of networking, a protocol can be considered as the envelope into which a message is placed, and the term describes the format and processes that the message must follow. The envelope and the data payload that is contained within are collectively known as a frame, and is the next layer of organization imposed on information being transferred between digital devices.

Figure 3.21 Routing around congestion. In this example network, communications between workstations A and B normally take place over the route A, E, F, B, since this is the shortest path between the two devices. If router E reports that it is busy, another route may be available, which provides a faster service, even though the route is longer.

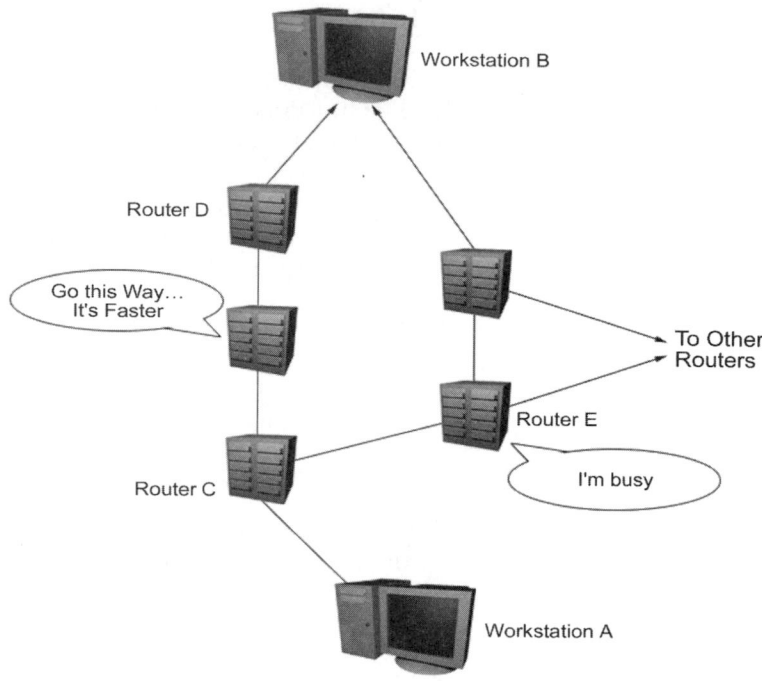

3.4.2 TCP/IP

TCP/IP is a useful protocol to explain, since it is the most prevalent protocol in use in the computer networking industry and the primary protocol in use on the Internet. It is simple to explain and contains all the major principles of internetworking. Explanations of TCP/IP lead to an understanding of what protocols in general set out to achieve, so other protocols can be understood in the context of TCP/IP.

Why TCP/IP?

The success of IP is largely due to the explosion of the Internet and the subsequent adoption of the protocol by corporates within their LANs and WANs.

TCP/IP stands for transmission control protocol (TCP) and Internet protocol (IP). This popular acronym encompasses two separate parts of the protocol suite, each performing a different function at different layers within the model. The entire protocol suite is made up of smaller parts, each performing a specific function or task.

The relationship between TCP and IP is best explained by using the metaphor recited by Vinton Cerf, who was one of the original

Advanced Research Projects Agency (ARPA) scientists experimenting with the concept of packet switching and is known to many as one of the fathers of the Internet, because of the part he played in turning the original ARPANET into the Internet. His explanation is adapted here (*webstory.com*, 1999):

Imagine a requirement to send a novel to someone using only postcards. To successfully send the whole novel, several thousand postcards may be required. Since the postal service is to be used as the transport mechanism (instead of Ethernet, token ring or some other transport mechanism), it is very likely that the postcards will arrive at the destination in the wrong order. Therefore, a system is applied whereby each postcard is numbered so that the correct order can be determined by the receiving party. This system of numbering is also useful since it is possible that some of the postcards will get lost during the journey.

In the case of lost postcards, the receiving party can request a particular postcard to be sent again, and in order for this to be successful, it is necessary for the sending party to retain copies of each postcard. In order to find out whether any postcards have gone missing, a system of acknowledgements is used, and these come back in the form of more postcards.

In building such a system, it is realized that if the sending mechanism is unreliable in one direction, then it is likely to be unreliable in the other direction, and so it is possible that some of the acknowledgements will also get lost. To compensate for lost acknowledgements, a system of timeouts is devised. If a particular acknowledgement is not received in a timely fashion, then a copy of the original is resent anyway.

This analogy neatly illustrates how TCP sits in the transport layer, on top of IP in the network layer, working together to create a more reliable medium.

Essentially, the Internet components and hardware that marshal this data from one place to another have a finite limit to the amount of work they can do. The limit is governed by processing power and memory and so it is generally understood that network components may need to drop packets if enough memory is not available during a particular burst of traffic. The QoS for the delivery of data on the Internet is no better than 'best endeavours' as a result.

In summary, TCP arbitrates the more basic IP in order to create a more reliable delivery mechanism.

Although the TCP/IP protocol was designed to scale to the size of the Internet, its open standard and simple bit-level routing mathematics made for a fast, tidy and obvious solution to the reliability conundrum.

TCP is the transport layer most commonly used with IP, but it is not the only one. Another common transport protocol used with IP is user datagram protocol (UDP) which is designed with less reliability and fuss in mind. UDP is noteworthy as it has specific applications for use in data streams over an unreliable network.

TCP requires acknowledgements for each packet sent, so that the delivery of data becomes reliable. However, this system of numbering and acknowledgements means that for delivering a real-time stream of data, there is a buffering activity in order to ensure that the entire stream is reassembled in the correct order before being transmitted onward to the software.

In other words, if a single packet containing data from within an audio stream is lost, then the rest of the data will be held up whilst this lost packet is found and replaced in the correct position before playback.

This process has an effect on the timeliness of a real-time stream. If audio were delivered over TCP/IP, there is a good chance that the result would be poor quality sound with delays, clicks, and pops where packets have not turned up in a timely fashion, and the D/A process has had to wait for the information to be delivered. UDP, on the other hand, does not require acknowledgements for each packet, and so would not invoke any further activity if certain packets were lost. This is important to streamed media such as audio since lost packets are of no value. Any software written to exploit this kind of behaviour would expect packets to be lost, and not interrupt playback to wait for lost packets to be replaced, instead carrying out some error correction routine (such as interpolation). Although this would result in loss of quality in an audio stream, certain defined audio performance requirements such as those defined as User Oriented and Consumer Broadcast are possible.

On the basis of UDP, it would be possible to send a stream of audio over a network, provided that the network was designed in such a way as to allow minimum disruption to the data flow.

Some additional criteria need to be considered in the overall network design, such as the amount of memory in routers, the rules under which IP packets are dropped, how many devices attach to any particular network segment, and when and how these are likely to be used.

How does IP work?

IP is a network layer or layer-3 protocol. It provides the actual addressing that allows packets to be routed to the correct destination. IP addresses are made up of four 8-bit bytes, called octets, and these are often written in decimal format for simplicity. To illustrate this, it is easy to see that this binary address notation:

```
10000111.10100101.00100011.11110000
```

is more easily written and understood as:

```
135.165.35.240
```

Another common annotation uses hexadecimal, and allows each octet to be represented in two halves, of 4 bits each. Using the example above, the last octet (11110000) can be more easily imagined as two nibbles, one of 1111 and the other of 0000. The binary equivalent of each is much easier to visualize since 1111 (2^4) is 16 in decimal, and 0000 is 0. The hex equivalent of 16 is F (from Chapter 1) and 0 is still 0, making the octet notation F0.

Using hex, the whole address would be written as:

```
87.A5.23.F0
```

IP addresses are organized into classes based on their first octet. Class A addresses begin with any number between 1 and 127, class B addresses begin with any number between 128 and 191, and class C addresses begin with any number between 192 and 223. Class D addresses are reserved for multicast addressing and begin with any number between 224 and 239. Class E addresses are also reserved and begin with any number between 240 and 247. Classes A, B, and C are the only publicly available addresses.

IP addressing and subnetting

As can be seen from the numbering scheme above, IP has a finite number of possible addresses, as there are only so many combinations to be had from four lots of 8 binary digits. The limitation is the number that can be represented by four 8-bit words, known as the address space. Since the Internet itself uses IP, most of the possible addresses have already been taken because of the huge number of computing devices attached to the Internet. Any organization wishing to attach their network to the Internet should ensure that an official IP address has been obtained. The only addresses left available are class C addresses (since classes D and E are reserved) and these are in limited supply.

In anticipation of this resource problem, the Internet Engineering Task Force (IETF – see Notes and further reading) began investigation into IPng (Internet protocol next generation) which resulted in IP version 6 (IPv6; Deering and Hinden, 1995), which increases the address space of IP, so extending the number of addresses available.

Addresses are handed out by InterNIC, which is a registered service trademark of the US Department of Commerce, and must be applied for before a computer can be successfully run on the Internet. There are several mechanisms, such as address translation, that can mask unofficial network numbers from the Internet, allowing organizations' computers to access the Internet without interfering with the official Internet addressing scheme, and without requiring a renumbering of the organizations internally assigned network numbers. In order to avoid these problems, most organizations rely on an Internet service provider, which already has a connection to the Internet that can be utilized for commercial purposes.

Design

A successful IP network relies on some management of the addressing scheme. The address scheme can be the application of some simple rules when setting up the network, or individual IP addresses can be automatically assigned when each device attaches to the network. IP addresses are assigned by a device attached to the network, known as a DHCP server. DHCP, or dynamic host configuration protocol, is a protocol which allows network administrators to manage centrally and automate the assignment of IP addresses in a private network.

The choice of address scheme should be made with a view of how the network is likely to grow and should be designed to be as flexible as possible. Addresses are critical because Internet packets (sometimes known as datagrams or frames) can be diverted through complex routes based on their address.

IP's four octet structure and binary mathematics make it ideal for structuring networks into groups, or subnets, and these correspond to the class structure of IP addressing.

When a class A address is assigned, this takes the form (in hex) of:

```
nn.hh.hh.hh
```

where n represents the network number, and h represents the host number. In this case, the organization owning the address may have up to 16 777 214 hosts on one network. Alternatively, the same organization may choose to subdivide the host addresses into separate networks and use the second octet as the internal network number, therefore creating 255 possible network numbers from the class A address. Since a class A address has been assigned, this freedom is available. When a class B address is assigned, the first two octets are given, and so the number is given as:

nn.nn.hh.hh

The number of possible hosts on a class B network is 65 536 (256 × 256). Once again it is possible to create subnets on this network, thus organizing it further. If a whole octet was used for the subnet numbers, then the number of possible subnets becomes 256 and the number of possible hosts on each subnet is reduced to 256.

A class C address is given as the first three octets so the number becomes:

nn.nn.nn.hh

and the number of possible hosts on this network drops to 255.

The idea of assigning addresses and network numbers is fundamental to the construction of a packet switched network, and so this is illustrated by the following example.

Example Studio owns a class B address, which always consists of the first two octets, and the address is 128.1.x.x. This number is assigned to the studio by InterNIC and as far as the outside world is concerned, 128.1 is a single network. However, within the studio, this number is divided up into subnets. The subnet is performed on the third octet, as shown in Figure 3.22, and reflects the network number that a particular computer is attached to. One particular computer of interest happens to be assigned the IP address of 128.1.4.3, and this computer sits on the network 128.1.4, and has the address on that subnet of 3.

Generally, subnets correspond to physical network segments such as separate Ethernet segments, or token rings. There are also perfectly good reasons to avoid segmentation of this kind, such as virtual subnets, where a group of disparately attached workstations are associated with each other.

Devices inside Example Studio contain information about the subnet structure in use. Therefore, once a packet destined for 128.1.4.3 arrives at the university, the routers immediately understand that the destination device is on network 128.1.4 and routes it to the studio Ethernet or token ring that has been assigned that subnet number.

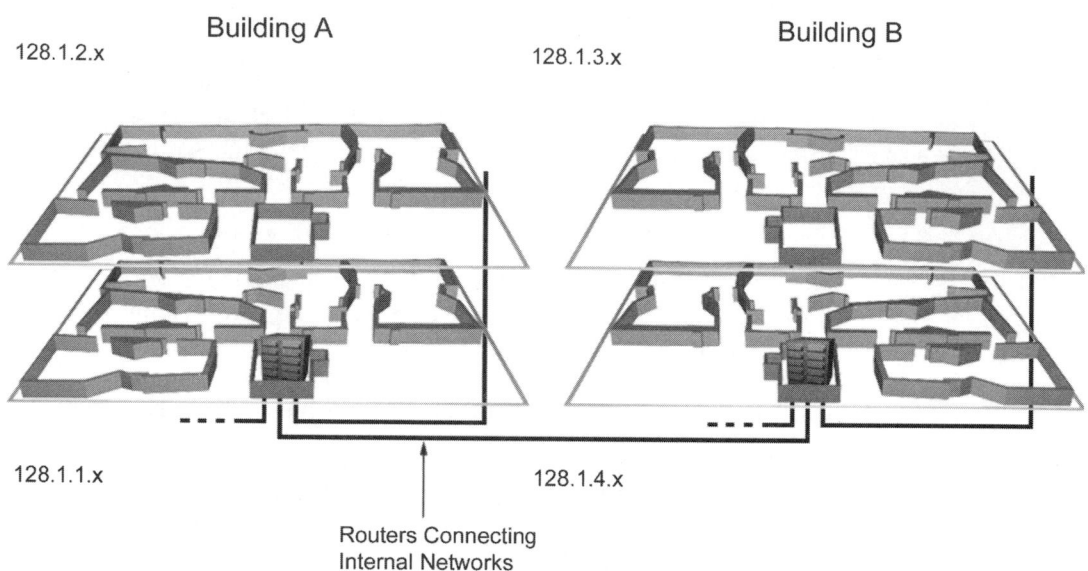

Building A   Building B

128.1.2.x   128.1.3.x

128.1.1.x   128.1.4.x

Routers Connecting
Internal Networks

Figure 3.22 Example Studio: subnet masks. In this example, the network is conveniently split so that it reflects the physical layout of the buildings. The machine room is located on network 128.1.1 and connected by a router to the other areas. The subnetting is performed on the routers.

Because TCP/IP is so prevalent on the Internet, it has also become common practice for commercial organizations to use IP on their own networks. For instance, in 1994 the British Standards institute specified that only IP packets would run on its networks.

The only public domain information that will be available to third parties is that 128.1 is the address for Example Studio. Unfortunately, this ability to add additional structure to the address via subnets was not present in the original TCP/IP specifications and so some older software is incapable of dealing with subnets.

Separating network numbers without using subnets has two disadvantages. The first and less serious is that it wastes address space. There are only 16 065 possible class B addresses and to assign ten of them to a single organization is wasteful, unless it is very large. The objection is lessened if three consecutive class C addresses are used for this purpose.

The second more serious problem is that assigning extra network numbers overloads the routing tables in the rest of the network. As mentioned previously, when a network number is divided into subnets, the division is known within the organization, but not outside it. Systems outside the organization need only one entry in their tables in order to be able to reach you.

Example Studio has one official network number assigned to it. Other universities wishing to exchange data only need the entry for 128.1 in the routing tables in order for the data exchange to be successful.

However, if the studio had several network numbers, then the division will be visible to the entire Internet. If 128.1 to 128.16 were used instead of subnetting 128.1, other organizations would need entries for each of those network numbers in their routing tables. The routing tables in many of the national networks have exceeded the size of the current routing technology at various times and it is considered wasteful for any organization to use more than one network number. This may not be a problem if a network is going to be completely self-contained and never attached to another network, or if only one small piece of it will be connected to the outside world. Nevertheless, most TCP/IP experts strongly recommend the use of subnets rather than multiple networks. The only reason for considering multiple networks is to deal with software that cannot handle subnets.

Allocation

Normally, deciding upon a subnet strategy is quite straightforward. Each physical network is assigned a separate subnet or network number. In some cases, it makes sense to assign several subnet numbers to a single physical network.

The Example Studio has a single Ethernet segment spanning three buildings using repeaters, illustrated in Figure 3.23. It is clear that as computers are added to this network segment, it will need to be split into several separate Ethernet segments. In order

Figure 3.23 Example Studio: subnet allocation.

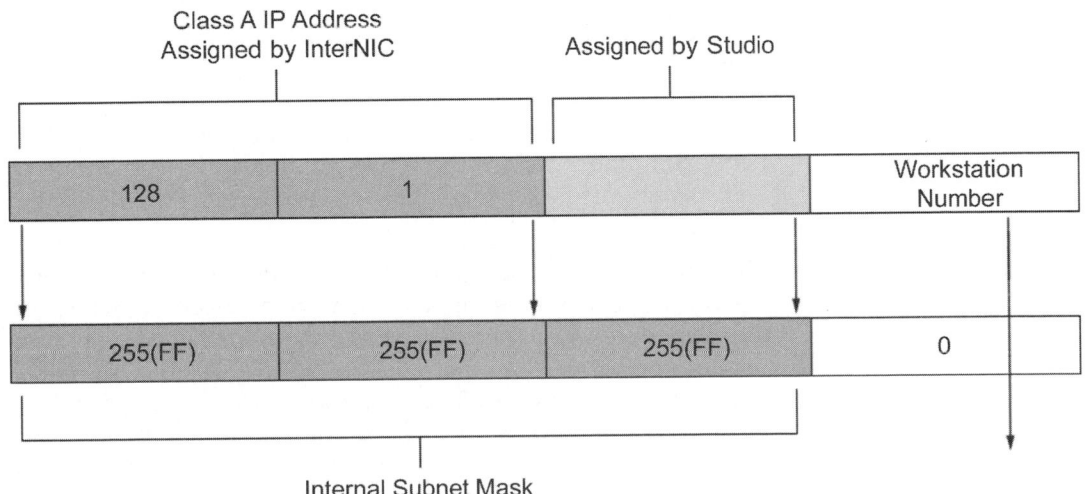

to avoid having to change addresses when this is done, three different subnet numbers are allocated to this Ethernet, one on each floor. Such a scheme is useful if there are no plans to split the segment, since it could be used to keep track of where computers are physically located.

From this it can be seen that IP addresses can be useful to identify the geographical location, purpose, department or any type of information that might be useful under different circumstances (device type, asset number, MAC card manufacturer and so on). The network designer may be able to incorporate some of this useful information into the IP addressing scheme. Networks are often addressed around the most significant numbering scheme within the organization (such as floor number). A subnet mask is used to separate the subnet from the rest of the address.

So far, the illustrations and examples have assumed that the first two octets are the network number, and the third octet is the subnet number. For class A networks, the first octet is the network number whereas for class B addresses, the standards specify that the first two octets are the network number. However, the owners of the address are free to choose the boundary between the subnet number and the rest of the address. Common practice is to have a one-octet subnet number, following the network address, so our network number now looks like this:

```
nn.nn.ss.hh
```

where *nn* still represents the network number (as assigned by InterNIC) and *hh* is still the host number; *ss* now represents the subnet number, which is not known to the outside world, but represents the internal network number where the host is attached.

Using this scheme, it is easy to see that there are 256 possible subnets and within each subnet there are 256 possible addresses. In fact, since it is bad practice to use the numbers 0 and 255 for network numbers (since the representation in binary are all 0s and 1s, respectively), then the possible number of hosts and subnets is more like 254.

Most physical layer transport mechanisms used in local area networks, such as Ethernet and token ring, specify a maximum number of devices that can be attached to the network and remain operationally supportable. In the case of Ethernet, this is considered to be around 30 stations for performance considerations, but is more dependent upon the physical cable length of the network bus. Therefore, it is unlikely that there will be more than 128 stations on any one subnet, although it is possible that a large and complex building such as an academic institute, governmental or large commercial organization may well have a network with more than 254 subnets on the whole network. In such a case, it is possible to define the subnet across the boundary of a single octet and define 10 bits (for instance) for the subnet number, leaving 6 bits for addresses within each subnet, as shown in Figure 3.24. The choice is expressed by a subnet mask, using 1s for the bits used by the network and subnet number, and 0s for the bits used for individual addresses. The subnet mask for Example Studio is given as 255.255.255.0, but if a 10-bit subnet scheme had been selected, the subnet mask would be 255.255.255.192, making the subnet mask appear in binary as follows (as shown in Figure 3.25):

```
11111111.11111111.11111111.11000000
```

Figure 3.24 Assigning multiple subnets to a single network.

The TCP/IP protocols allow for computers to send a query asking what the subnet mask is. If the network supports

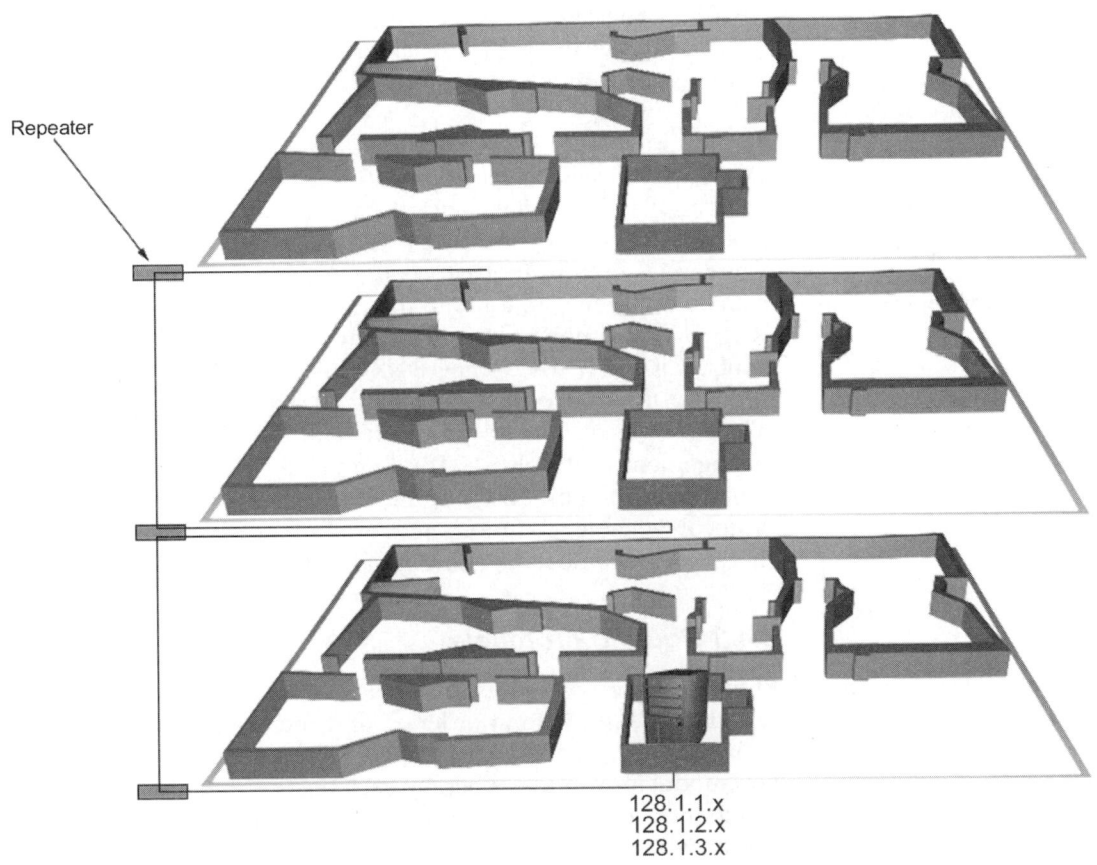

Repeater

128.1.1.x
128.1.2.x
128.1.3.x

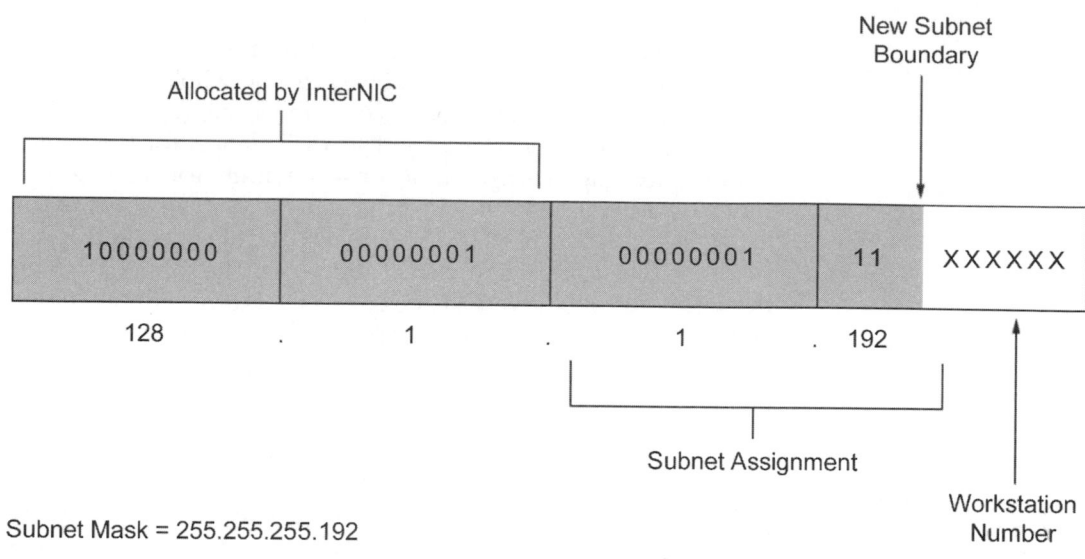

Subnet Mask = 255.255.255.192

Figure 3.25 Non-boundary subnetting. The three leftmost bit positions within the last octet are part of the network number assigned by the organization. This technique is used when it is unlikely that the number of workstations attached to a single network number will exceed the number represented by the remaining bit positions.

broadcast queries, and at least one computer or gateway on the network knows the subnet mask, it may be unnecessary to set it on the other computers.

Applications

IP is an efficient layer-3 protocol used for many types of applications. This efficiency allowed IP to function over low bandwidth wide area network (WAN) links, which led many commercial organizations to connect their sites using TCP/IP even before they connected to the Internet. Some of the common TCP/IP application protocols include file transfer protocol (FTP), Telnet, domain name system (DNS), simple mail transfer protocol (SMTP), multipurpose Internet mail extensions (MIME), X Window system, remote procedure call (RPC), simple network management protocol (SNMP), line printer Daemon (LPD), and dynamic host configuration protocol (DHCP). These application protocols enable common network applications such as file sharing, electronic mail, printing, and device management.

3.4.3 Routing protocols

Routing is the process used by networks to determine the path that data should take in order to find their destination and is a network layer function. Packet switching networks use a hop-by-hop model for routing, whereby each host or router examines the destination address in the IP packet header and then determines the next hop that will bring the packet one step closer to its destination. The process is repeated for each step along the pathway. There are two mechanisms required to make this work. The first is a table of addresses that are known to the router that can be used to match destination addresses with the next hop. The second function required to make this happen is a protocol that the routers themselves can use to exchange information within the tables regarding the status of links or routes.

A popular phrase of network engineers is 'real network engineers construct routing tables by hand', but routing protocols are available to automate the job of inspecting routes and these work on the network, constantly updating the routing tables. Routing protocols are the core of the hacker's Internet, because it is here that all the decisions about connections and interconnections get made.

Network engineers assign costs to network paths, and routing protocols select the least-cost paths to the destination, usually based on time and hop count.

To make a real-life example, routing protocols compare to free-market economies. In both systems, a large group of nodes (or companies) base their decisions on some form of cost effective-ness. The end result is a reasonably efficient, self-adjusting distribution of resources, either bandwidth or goods.

A router, like any commercial business, computes its cost and adds on some profit for its part in the transaction. In the case of routing, the additional cost is one extra hop for each node in the pathway.

The two main types of routing are distance vector routing and link state routing. Each type of routing has its favourite protocol to fulfil the requirements. With distance vector routing, the protocol is called the routing information protocol (RIP), and for link state routing, the most popular protocol is open shortest path first (OSPF).

Distance vector routing

Distance vector (DV) routing protocols require each router to inform its neighbours of its routing table. For each network path, the receiving routers pick the neighbour advertising the lowest cost and then add this entry into its routing table for re-advertisement. 'Hello' and 'RIP' are common distance vector routing protocols. Common enhancements to DV algorithms include split horizon, poison reverse, triggered updates, and holddown (*RIP protocol specification* – see Notes and further reading).

RIP protocol overview

RIP is probably the most widely used internal Internet routing protocol and is a distance vector protocol based on a 1970s Xerox design, again from the PARC research establishment. The protocol was ported to TCP/IP when LANs first appeared in the early 1980s and has hardly changed since. RIP suffers from several limitations, some of which have been overcome with RIP-2.

RIP uses 4 bits to count router hops to a destination and therefore RIP cannot support networks wider than 15 hops (so 16 is interpreted as infinity). RIP also has no direct subnet support, since it was originally deployed before subnetting became popular. RIP can be used on networks with subnets, although there are some limitations to its use in these environments. Probably the biggest problem with RIP is that every 30 s or so, a RIP router will broadcast lists of all the networks and subnets it can reach. Depending on the lengths of these lists, which depend on the size of the network, bandwidth usage can become prohibitive, especially on slow links.

Despite these problems, RIP also has several benefits. It is in widespread use and is the only interior gateway protocol that can be counted on to run everywhere. Additionally, configuring an RIP system requires little effort beyond setting path costs, and this only needs to be done if there is a significant reason for doing so, since by default the RIP protocol will report the cost of each hop as 1. Finally, RIP is a simple protocol at the mathematical level, and so it does not incur much processor overhead on the routers.

Link state

Link state routing protocols require each router to maintain at least a partial map of the network. When the state of a network link changes, by becoming unavailable or available, a notification packet called a link state advertisement (LSA) is flooded throughout the network. All the routers on the network note the change, and adjust their routing tables accordingly. This method is considered to be more reliable, easier to debug and less bandwidth-intensive than distance vector routing. However, it is also more mathematically complex and therefore more processor intensive.

OSPF-2 protocol overview

Open shortest path first (OSPF) is a more recent entry into the routing scene. Documented in RFC1583 it was originally sanctioned by the IETF for the intention of becoming the Internet's preferred interior routing protocol. It has a complex set of options and features, not all of which are available on all implementations. The protocol is specifically designed to scale up to larger networks and has full subnet support. OSPF uses small 'hello' packets to verify link operation without transferring large tables. In stable networks, large updates occur only once every 30 min.

OSPF is of special interest to time sensitive applications, since it is the first routing protocol to be able to route packets based on the type of service (TOS) they require, and this information is held in the TOS field within the packet. This functionality requires co-operative applications throughout the network in order to be effective.

OSPF has some disadvantages, such as the complexity of the protocol and the demands on memory and computation that it makes. Although link state protocols are not difficult to understand, OSPF muddles the picture with plenty of options and features.

OSPF divides networks into several classes, including point to point, multi-access, and non-broadcast multi-access. A single serial link connecting two routers together would be a point-to-point link, whilst an Ethernet or token ring segment would be a multi-access link.

3.4.4 Conclusion to protocols

There are a great many protocols in use throughout the networks of the world, and it is common for these protocols to be adapted in some way, either by specification, or by use of a gateway.

TCP/IP was chosen to illustrate the functionality of protocols simply because it is an open and widely understood technology, used everyday by people connecting to the Internet. Its application and theory go far beyond what has been covered here and the interested reader is directed to the references and the many thousands of sources in print and on the Internet.

Recommended for the reader who is a little frightened of the language used in these reference sources is the excellent '. . . for Dummies' series, which includes titles on both the Internet and TCP/IP. Far from being 'for Dummies', these books cover a complete initiation of the chosen subject written in plain English.

Notes and further reading

Deering, S. and Hinden, R. (1995) *IETF RFC 1883 Internet Protocol version 6 (IPv6) specification.*

Electronic Industries Alliance. 2500 Wilson Boulevard, Arlington, VA 22201-3834, USA. http://www.eia.org.

FDDI Basics. Bay Networks Technical Response Center. Technical Tip TT-FDDI-9508001.

IETF Secretariat. c/o Corporation for National Research Initiatives, 1895 Preston White Drive, Suite 100, Reston, VA 20191-5434, USA.

RIP protocol specification, RFC 1058. For discussion of distance-vector or Bellman-Ford algorithms.

Telecommunications Industry Association. 2500 Wilson Blvd, Suite 300, Arlington, VA 22201 USA. http://www.tiaonline.org.

The Institute of Electrical and Electronics Engineers, Inc. IEEE Operations Center, 445 Hoes Lane, Piscataway, NJ, USA.

US Department of Commerce. Constitution Avenue, Washington DC, USA. http://www.internic.net.

webstory.com – yesterday. Courtesy of the BBC. Broadcast 1999.

WWWebster Dictionary (2000) Merriam-Webster Inc., PO Box 281, Springfield, MA 01102. http://www.m-w.com/.

4 Audio interfaces

4.1 Introduction to audio interfaces

In the previous chapter, practical examples were given for each of the layers concerned with the transport of data from one place to another. Many of the examples are related directly to the audio studio, such as those describing fibre and other types of cables. IP is also used within the audio industry, for instance in the AES24 recommendation, which details the control of sound systems over computer networks and is discussed in more detail later.

This chapter presents and explains the most common technologies and mechanisms used specifically for the transfer of audio data, with particular emphasis on those candidates used for delivering QoS for streams of data, as described for performance oriented. Such interfaces are already prevalent within the audio industry. Where possible, these are compared to the 7-layer model for reference to the computer-networking counterpart discussed in Chapter 3.

The technologies within this chapter are presented with the intention of exposing the inner mechanisms for the purposes of utilization and comparison with other communications technologies.

4.2 Audio interfaces

As concluded in Chapter 1, standards play an important role in most technical fields, including audio and digital audio. Many of the audio interfaces presented in this chapter are based on a core set of standards, but contain enough differences such that two apparently similar products will not interconnect directly. For instance, the Sony and Philips initiative and the Audio Engineering Society (AES) produced related works, with the S/PDIF standard now considered to be a consumer deployment of the harmonizing digital audio standard IEC 958.

4.2.1 Topology

Professional digital audio equipment used in some recording studios is connected in a manner that reflects the unidirectional flow of signal towards an endpoint, such as a tape device as shown in the configuration in Figure 4.1. There may be no provision for data to be carried in the opposite direction between devices. Instead, the return path takes an alternative route through equipment, towards a new endpoint, the loudspeakers. Multiple destinations may be envisaged in the event that surround systems deliver digital information to the loudspeaker,

Figure 4.1 Schematic-style diagram of signal path to multiple endpoints. Audio signal effectors connected together form a network with an intended direction to the signal flow towards the endpoint, such as loudspeakers (shown) or recording device.

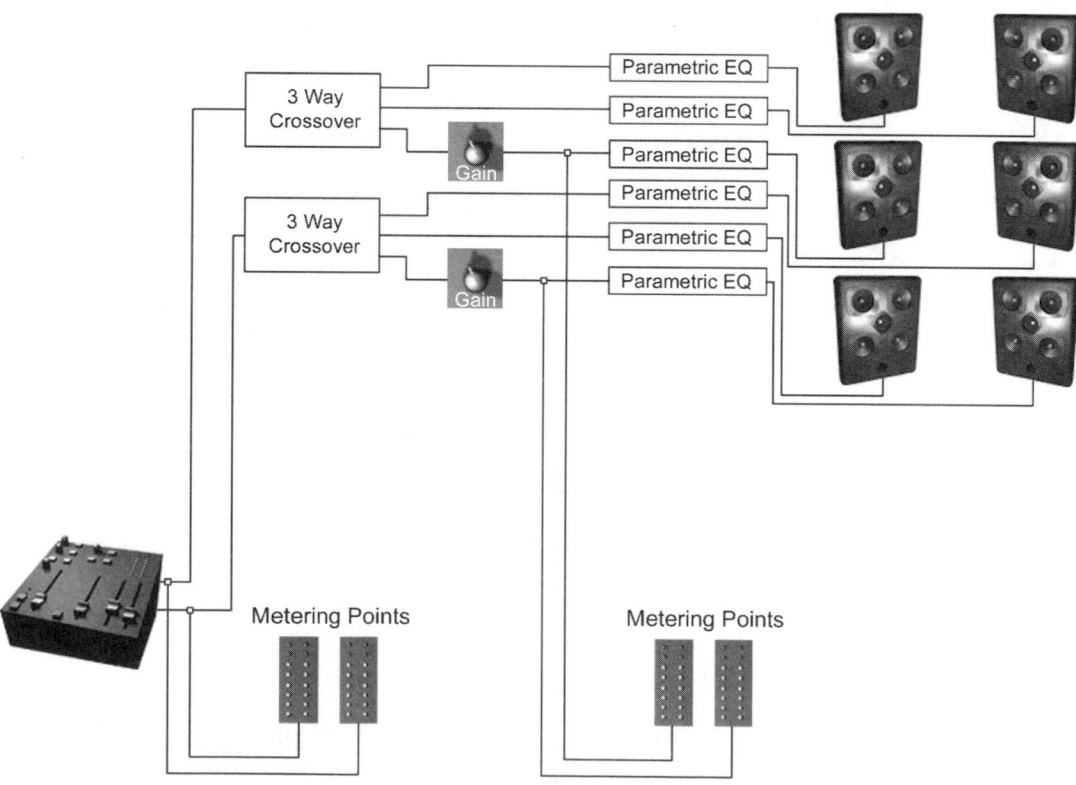

or recording equipment is added for instance. Added to this is the necessity to monitor and change the assignments in small percentages during any mixing process, perhaps in a repeatable performance by recording control data.

To support a studio-related operation in this manner, the data must also remain synchronized throughout the entire process, whilst ideally maintaining as little delay as possible.

At a simpler level, the routing of audio through various paths, such as inside a mixing console, can be likened to routing over links in a network. The biggest difference when cabling audio connections compared to a network is that there is very much a direction to the audio, depending upon the connection of inputs and outputs. The engineer or operator makes adjustments to the audio signal's destination within the console. An example of this is routing a single audio signal through successive buses, or to route the signal to some other processing such as a compressor or delay, via an effects send and return as shown in Figure 4.2.

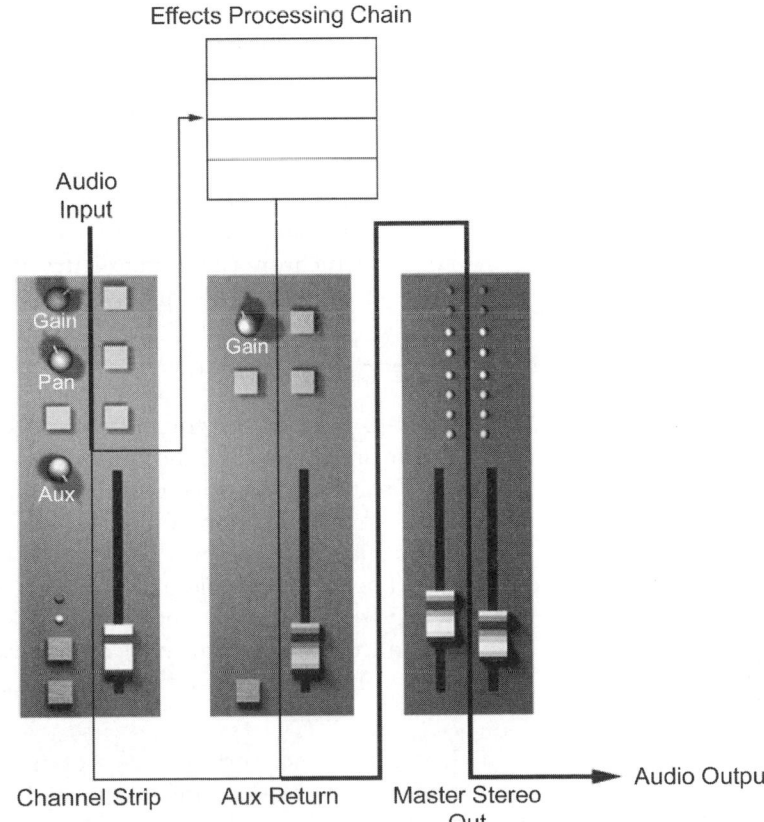

Figure 4.2 Simplistic representation of audio routing assignments, showing connections made inside a console. In this configuration, channels assigning output to the auxiliary bus send the signal to an external effects rack before returning to the auxiliary return, and on to the stereo master output.

This is as opposed to computer networks where devices broadcast data directly onto a shared medium, and information contained within the transmission itself determines the destination.

Audio devices within professional audio studios are usually connected together directly, using a single dedicated cable between each device in a point to point topology.

The most common interfaces that make any external digital connections within Figure 4.2 describe point to point protocols. Several devices may be connected, although this is achieved by daisy chaining through further machines and each device receives the audio data in turn. As such, audio interfaces are designed to transfer data in real-time streams, with no QoS requirement for multicasting or destination information contained within the data stream. This differs from the QoS of computer networks, which developed with the efficiency of communication channels in mind. Computer networks are designed principally for the purpose of file transfer, with a QoS that is not time critical in nature.

The fundamental difference between the two types of interface is described by the difference in QoS. The area between the two is explored more fully in references to voiceover IP (Held, 1998), Internet radio, the Internet2 project, and IPv6. The discussion specifically concerns the ability of computer network technology, particularly the Internet, to deliver real-time multimedia information. It is also worth considering the QoS for other data types such as synchronization and control data as well as digital audio data, since these could also utilize simplified interconnection.

4.2.2 Cable strategy

Point to point devices do not normally share a cable and attempt to transmit onto it at the same time. Instead, the connection is unidirectional. Unidirectional (or simplex) means that data only travel in one direction down the cable. Instead, the point to point connection employs a second communications channel, which might be an extra cable pair or fibre where an RX/TX connection achieves two-way communication. RX and TX cables are identical, since the TX (transmit) of one device connects to the RX (receive) of the other device, and represents input and output of the data stream.

In computer networks, the receipt of data is checked for errors so that lost information can be requested and retransmitted. In digital audio point to point connections, there is a minimum mechanism used, since the protocols, interfaces, devices, and topology are all built to service the QoS requirements of streamed media.

There are a number of types of digital interface in use throughout the professional and consumer audio worlds. Some of these are standards, and some of these are proprietary, or manufacturer specific. All of those in common use are designed to carry audio most often encoded into 44.1 kHz 16-bit word length at the very least. Most are also designed to carry other resolutions, including 32 kHz, 48 kHz, 96 kHz and up to 24-bit word length. Most carry streams of stereo information, and some can carry several channels at once.

It is not intended to detail the exact specifications of all the standards here, and descriptions of the interfaces are deliberately limited to outline information. The interested reader is directed to *The Digital Interface Handbook* by Rumsey and Watkinson and the further reading section.

4.2.3 Audio standards

Within the audio industry, the Audio Engineering Society has led the way in determining digital audio transfer standards. The AES is a professional society and not a standards body; its committees are accredited by ANSI and its recommendations have been the basis for many standards, especially those involving two-channel audio interfaces. Standards bodies such as ANSI, BSi, EIAJ (Electronic Industries Association of Japan) and IEC (International Electrotechnical Commission) administrate and co-ordinate the opinions of all interested parties when creating a standard.

The AES first published its viewpoints on two-channel digital transfer in the document AES3–1985. This work was undertaken to cover professional applications, and parallel work on a consumer version performed by Sony and Philips resulted in the S/PDIF interface for the CD system slightly earlier, in 1984.

Many of the standards bodies produced standards based on the AES3–1985 recommendations and each body assigned its own document identity to the standard. Therefore, the same standard may be called by a number of different names. Furthermore, although the standards are very similar there are enough differences to ensure that interconnection cannot always be assumed.

One of the important concepts to understand is that the format of the data is not necessarily related to the electrical characteristics of the interface as explained by the modularity of the 7-layer model in Chapter 2.

4.2.4 AES/EBU

The AES/EBU interface is perhaps one of the most popular standards in use in the audio facility and is described almost identically in four documents: AES–1992, IEC 958 (Type 1), CCIR Rec. 647 and also in EBU Tech.3250E.

The standard allows for two channels of digital audio to be transferred over a balanced interface. The interface specifies serial transmission of data, and the clock signal is contained within that data.

At the physical layer of the 7-layer model, the interface can use a range of media types, such as balanced or unbalanced, optical and coaxial cables.

Balanced interface

All the standards referring to professional or broadcast use specify a balanced electrical interface conforming to CCITT Rec. V.11, and there are similarities between this specification and the RS422A standard, but although RS422 drivers are used in many cases, they are not identical.

A $110\,\Omega$ input impedance was originally recommended for output, cable, and input. Later revisions changed the wording slightly to recommended that the impedance value be the same as the transmitter and the transmission line.

The standard specifies audio XLR-3 connectors (IEC 60268-12 – see Notes and further reading), as used commonly in audio studios and illustrated in Figure 4.3.

Pin 1 is the shield and pins 2 and 3 are the balanced data signal. The polarity is not essential, although manufacturers generally stick to the convention that pin 2 is assigned as '+' and pin 3 is '–'.

Figure 4.3 IEC 60268-12 connector (also known as XLR or AES/EBU electrical connectors).

4.2.5 S/PDIF and IEC 958 type 2

S/PDIF is becoming more common in equipment with higher specification due to market forces. As the price of digital electronics falls, or more accurately as more digital power is offered for the same money, so it has become cheaper to offer digital audio manipulation that was previously only available to the higher budgets of the professional audio world. These types of electronics include sound cards for PCs which contain digital signal processors (DSP) designed for the manipulation and routing of audio. The kind of equipment available now for a few hundred pounds can match the functionality of equipment previously sold for several thousands of pounds, and follows the general trend of technology products.

Standards and administration

The AES along with the European Broadcast Union (EBU) produced almost identical documents administrated together by the International Electrotechnical Commission (IEC).

The IEC is a worldwide organization for standardization comprising national electrotechnical committees. The object of the IEC is to promote international co-operation on all questions concerning standardization in the electrical and electronic fields. The IEC collaborates closely with the International Organization for Standardization (ISO) in accordance with conditions determined by agreement between the two organizations.

The consumer interface developed by Sony and Philips (S/PDIF) coincides with the AES3–1985 recommendations and IEC 958 is closely related to the S/PDIF interface. The main difference between the professional and consumer standard is that the consumer interface uses an unbalanced connection and recommends an impedance of 75 Ω. The standard does not specifically mention that unbalanced connections should be used within consumer products.

An optical interface is also available as a possible medium for transmission and is mentioned as 'under consideration' within the original IEC 958 committee documents. Since then the standard has been completely reviewed and renumbered accordingly, with the latest revision split into several parts. The new publication number is IEC 60958. This has four parts, three of which were published in 1999.

Part 1 describes a serial, simplex, self-clocking interface for the interconnection of digital audio equipment for consumer and professional applications, using linear PCM coded audio samples. It provides the basic structure of the interface (IEC 60958-1 – see Notes and further reading).

The second part is entitled IEC/TR3 60958-2 (1994-07) Digital audio interface – Part 2: Software information delivery mode. This part proposes a software information delivery mode to be used for the transmission of basic software information. This is a transmission format that uses a channel status bit, assigning mode bits 6 and 7 as identification bits.

Part 3 describes consumer applications and Part 4 describes professional applications intended for use with shielded twisted-pair cables over distances of up to 100 m.

Physical media

The standard was designed as an interface for the transfer of digital audio data to be included on consumer products such as CD players and DAT machines. Implementations of the unbalanced interface use RCA phono connectors (Figure 4.4) and coaxial cable, although optical fibre is also used on some audiophile equipment. Media filters are available to convert from fibre to coaxial cable or vice versa.

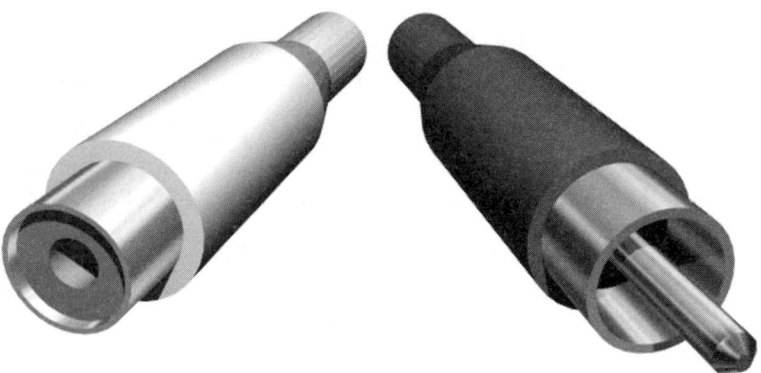

Figure 4.4 RCA connectors.

Optical interface

The generally adopted optical interface is described within the document EIAJ CP-340 and consists of a transmitter using a wavelength of 660 ± 30 nm with a power of between –15 dBm and –21 dBm. Receivers complying with the specification should correctly interpret the data down to –27 dBm.

The preferred connector is specified in EIAJ RCZ-6901 and is shown in Figure 4.5.

Figure 4.5 EIAJ RCZ-6901 AES/EBU preferred optical connector. One connection is required for each direction of transmission – transmit and receive (TX/RX).

This interface is usually found in consumer products such as DAT recorders, sound cards (and other PC peripherals), audiophile standalone D/A converters and CD players.

The transmission mechanism is normally a light emitting diode (LED) which is cheaper to implement than the laser version. The receiving device contains a photodetector. The TOSLink interface from Toshiba is a popular implementation using a 0 to 5 volt source, with an identical data structure to the electrical interface.

Coaxial interface

The possibilities for transferring AES3 data over coaxial cable are discussed in the AES SC-02-02 Working Group on Digital Input/Output Interfacing (Meeting, May 1999), where it is acknowledged that the provisional document AES-3id–1995 is very similar to SMPTE 276M–1995. This document specifies a $75\,\Omega$ video-like coaxial cable to carry audio signals over distances of 1000 m. The signal level is similar to that of video at around 1 volt, although the data structure remains the same as AES3.

Initial tests of the format were successful in transmitting an AES3 digital audio signal over 1300 km without any noticeable corruption.

Data link

The lack of any requirement to account for shared access to the transmission medium within these interfaces results in a far simpler data link layer for audio interfaces.

Encoding

When data are placed onto the cable, time slots 4 to 31 are encoded in bi-phase mark scheme. The Manchester encoding mechanism is an example of mark coding, where a clock is incorporated with the data, thus having the advantage of ensuring the correct understanding of bit timing and boundaries during the transmit/receive process, which is known as clock recovery.

Each bit is represented by a symbol comprising two consecutive binary states; the first state of a symbol is always different from the second state of the previous symbol. The second state of the symbol is identical to the first, if the bit to be transmitted is logical 0. However, it is different if the bit is a 1 (see Figure 4.6).

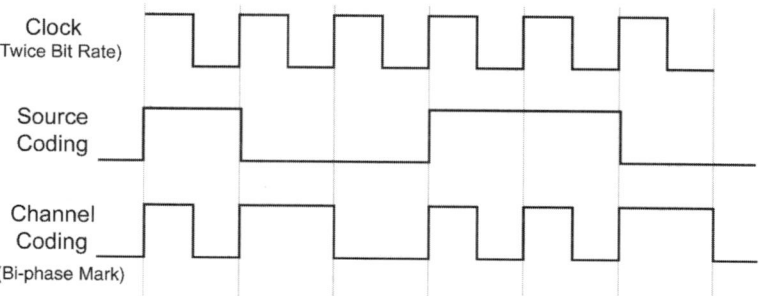

Figure 4.6 Bi-phase mark coding implementation in IEC 60958.

The original version of the standard from 1985 specified a voltage of between 3 and 10 volts but the standard was changed in order to conform more closely to the practice of many manufacturers of connecting an RS-422 driver directly between the two legs of the source.

Framing

Audio data are sampled and each sample is then placed within a word of a fixed length using the PCM mechanism. The recommendations describe two samples, one from each of the two stereo channels, placed within a subframe and transmitted over one sample period.

The function of the subframe is to indicate the start and end of each sample. The starting signature of the subframe consists of one of three patterns, which deliberately breaks the rules of bi-phase mark coding, in order to make it easily identifiable by the receiving device. A further 4 bits of additional data are also carried within the subframe, and this extra space can be used for a number of purposes such as a buffering zone, in case increased word lengths are used during the sample phase. Other information contained within the subframe is the validation bit, channel status, a user assignable bit and a parity bit.

Some flexibility is accounted for within the standard, allowing payloads of 20 and 24 bits or fewer. A subframe consists of 32 time slots, numbered from 0 to 31 as shown in Figure 4.7.

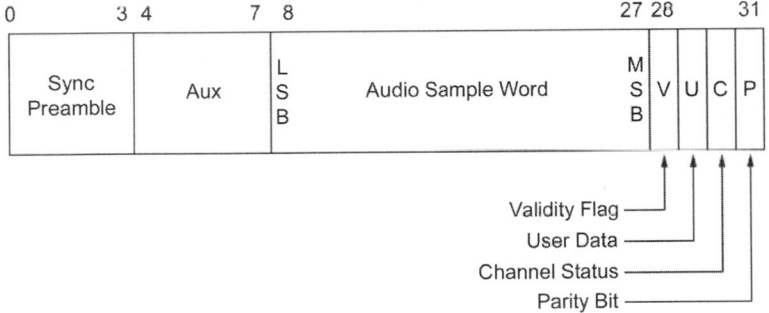

Figure 4.7 IEC 60958 subframe format. Source: IEC 60958-1 *Digital Audio Interface – Part 1: General* (1999).

Slots 0 to 3 carry one of the three permitted preambles and slots 4 to 27 carry the audio sample word payload. The MSB is designated as slot 27.

When a 24-bit coding range is used, the LSB is located in slot 4 and when a 20-bit coding range is used, the LSB is located in time slot 8, with time slots 8 to 27 carrying the audio sample word. Slots 4 to 7 are then left for other applications, and known as auxiliary sample bits. If the source provides too few bits, the auxiliary bits are set to 0.

Error checking

The parity bit is set such that the number of 1s within a word is always even, and in the event that a parity error is detected, an error handling technique is invoked. In the event of an error burst, where multiple parity errors are detected, muting is applied.

Information and instruction are encoded into the stream using the channel status bit, which is removed by the receiving device and stored in consecutive order to create a 24-byte word, every

192 frames. Each binary bit or group of bits within this 24-byte word has a specific meaning relating to, for instance, the negotiation of sample rates, or some other interface instruction or operation.

Channel status

Apart from the difference in physical interconnections, another key difference between consumer and professional interfaces is mentioned in the context of the channel status bits. The data format of the subframe specified for the consumer interface is the same as that used in the professional format, although the use of the channel status bit is significantly different. The second byte in the consumer interface is set aside for category codes, which are used to determine the type of consumer usage. The user bits of the subframe carry information such as track identification and cue point, and ensure that the track start ID is transferred along with the audio data.

4.2.6 AES10–1991, standard multi-channel interface

In 1991, a number of digital audio manufacturers combined to propose a multi-channel serial interface with the working title of AES10–1991. The interface described within the document became known as MADI (multi-channel audio digital interface). The MADI interface is also based on the AES3–1985 recommendations and is designed to be transparent to the AES/EBU data.

General description

MADI allows for up to 56 channels of audio information, and has applications within large-scale digital routing systems and the interconnection of multi-channel audio equipment. MADI uses serial transfer of information and a much higher transfer rate in order to move the increased amount of data around.

Physical

MADI is intended to be asynchronous in nature and therefore devices are locked to a common clock signal, which uses the same reference signal specified within the AES/EBU recommendations.

A new recommendation within MADI, required because of the increased amount of data, is that the data rate is specified as 125 MB/s, regardless of the sampling rate or number of channels.

The maximum cable length between two MADI devices is specified at 50 m, although longer distances can be achieved by using fibre as the interconnecting medium.

Frames

In order to achieve similarity with the AES3 recommendations, AES10 uses the same basic subframe structure with either 20- or 24-bit audio data, along with the same status bit structure from AES3. In order to transfer multiple channels, there are some important differences that are worth detailing.

The first 4 bits of the subframe do not break the rules of bi-phase mark coding in order to mark the start of a frame as in the AES recommendations. Instead, the bits are used as header information. Additionally, the AES/EBU frames are linked together to form a superframe containing up to 56 AES/EBU type subframes (Figure 4.8).

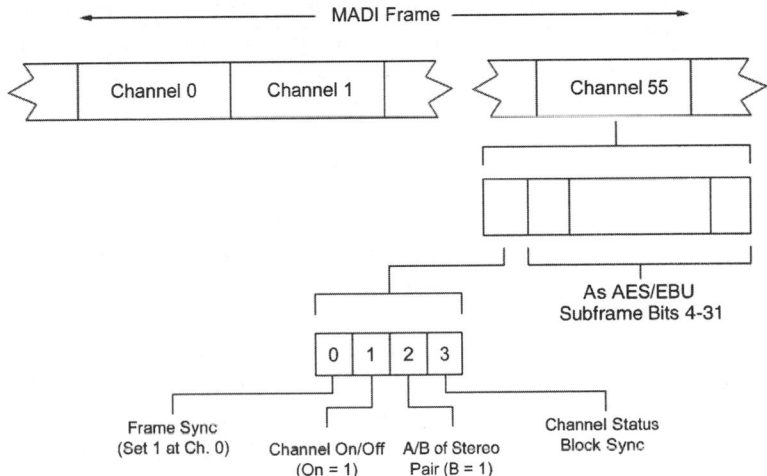

Figure 4.8 MADI super-frame, showing time-division multiplexing by channel. Source: Rumsey, Francis and Watkinson, John (1995) *The Digital Interface Handbook*, Second Edition, Focal Press, p. 127.

As per AES3, audio is sampled using the PCM sampling method and placed within a fixed length word. Within the MADI super-frame, audio samples must not use different sampling rates, and so when a change to one channel's sample rate is required, MADI enabled devices change the sample rate for all channels at once, so retaining synchronization.

Addressing

The word is surrounded by the AES3 subframe structure, and each of the 56 subframes is placed side by side to create the super-frame. Audio channels are correctly identified within the super-frame by using this simple ordering process. As such, a primitive addressing scheme is achieved without the need for address space within the bit structure of the frame.

The regularity requires a constant data rate within the communication channel, and a padding mechanism is added to fill any unused gaps. The regularity in the frame structure enables the MADI frame structure to be used in audio routing applications. By using this stable structure, multi-channel matrix mixers and routers have been developed to direct the audio channels to the correct destination (see the Studer, Pro-Bel, etc. product ranges).

4.3 Control data

The difference between digital audio data, MIDI, synchronization and any other kind of control data is a distinct one. Digital audio data contain audio information which, when decoded, can be detected by the ear. Control data, however, are simply data designed to control a device. As mentioned in Chapter 1, the device may be a toy car or a musical instrument, and the control data used to control it are inevitably designed specifically for that purpose.

4.3.1 MIDI

The musical instrument digital interface (MIDI) protocol provides a standardized and efficient means of conveying musical performance information as electronic data, which has been in use for some time. MIDI information is transmitted in 'MIDI messages', which instruct a music synthesizer or sound module on how to perform a piece of music, comparable to the human task of reading sheet music. The synthesizer receiving the MIDI data generates the actual audio.

General description

The MIDI interface on a MIDI enabled device will generally include three different MIDI connectors, labelled IN, OUT, and THRU. The devices are connected in a chain, with the OUT or THRU port of the first device connected to the IN port of the second device. The OUT or THRU port of the second device in the chain will be connected to the IN port of the third device and so on. The MIDI data stream originates from a source, such as a musical instrument keyboard, or MIDI sequencer. A MIDI sequencer is a device which allows MIDI data sequences to be captured and stored by saving files on, for example, a computer. The files can then be edited, combined, and replayed. Commonly this is implemented within software on a PC, although dedicated sequencing devices are also available. Figure 4.9 shows a simple MIDI system, consisting of a MIDI sequencer and/or MIDI sound modules. It should be noted that a sequencer or DAW

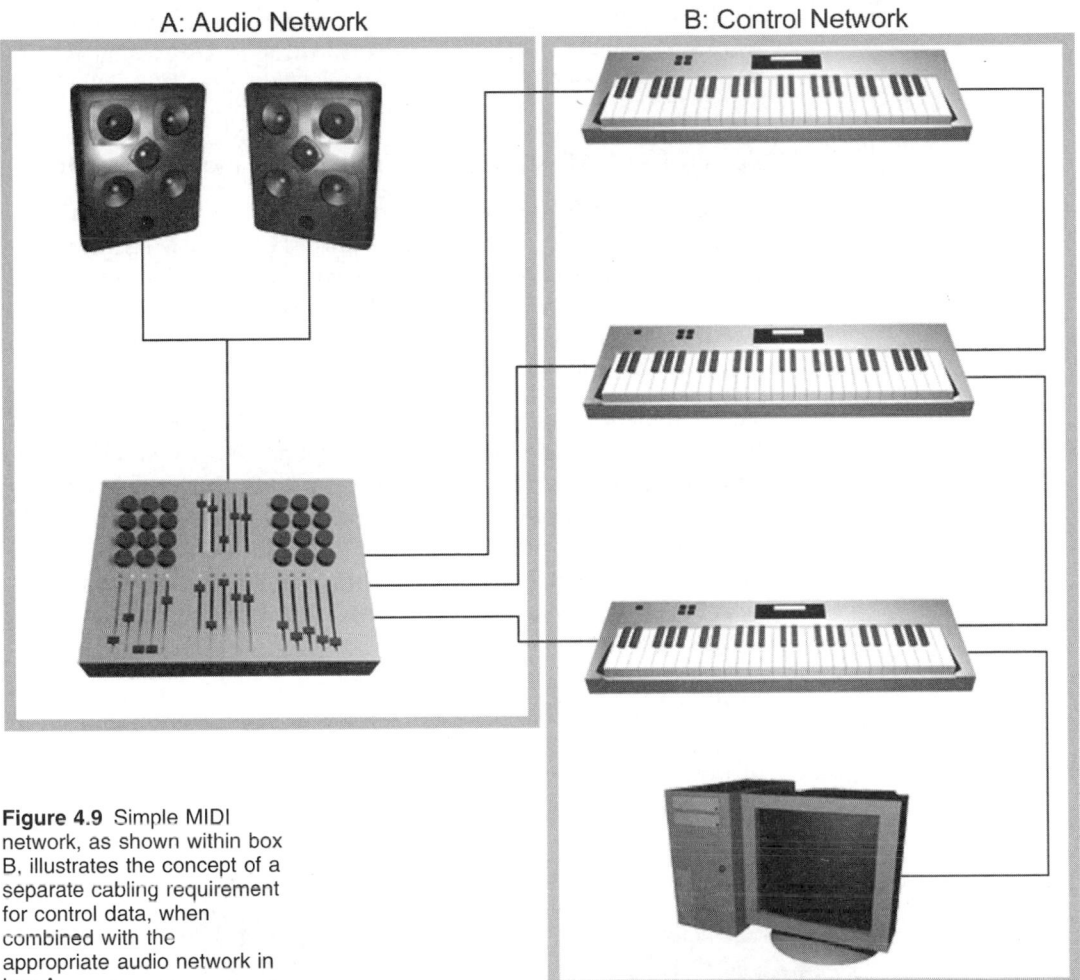

Figure 4.9 Simple MIDI network, as shown within box B, illustrates the concept of a separate cabling requirement for control data, when combined with the appropriate audio network in box A.

could replace the device represented as a workstation. In the case of a DAW, the cabling is further confused as the audio network returns to this machine, creating a loop connection between control and audio networks.

Standards and administration

The MIDI 1.0 Detailed Specification provides a complete description of the MIDI protocol.

Although the MIDI Specification is still called MIDI 1.0, the original specification was written in 1984 and there have been many enhancements and updates to the document since. Besides the addition of new MIDI messages such as the MIDI Machine

Control and MIDI Show Control messages, there have also been improvements to the basic protocol, adding features such as Bank Select, All Sound Off, and many other new controller commands.

Until 1995, five separate documents covered basic MIDI, additions (MSC & MMC), Standard MIDI Files and General MIDI.

In January 1995, the latest versions of these documents were compiled together into the 95.1 version. The basic MIDI specification that was used within the 95.1 compilation was version 4.2, which was a compilation of the Detailed Specification v4.2 document and the 4.2 Addendum. Version 95.1 integrated the existing documents and fixed some minor errors.

The MIDI Manufacturers Association was formed in 1984 as a trust to maintain the MIDI specification as an open standard and provides forums for discussion of proposals aimed at improving and standardizing the capabilities of MIDI-related products. The MMA provides a process for adoption and subsequent publication of any enhancements or clarifications resulting from these activities (MIDI Manufacturers Association Inc. – see Notes and further reading).

Physical interface

According to the MIDI 1.0 Specification, the only approved physical connector is a 5-pin DIN plug as shown along with the pin designations in Figure 4.10. It is also possible to send MIDI

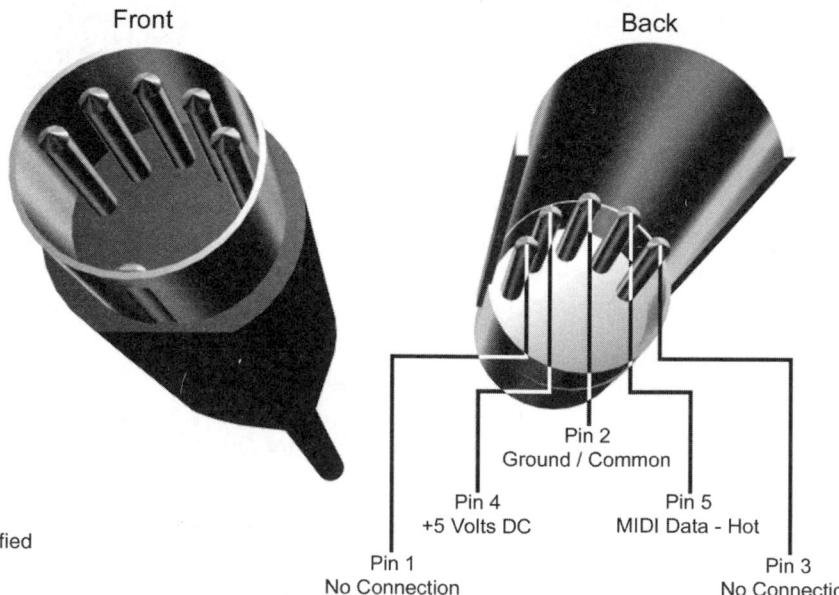

Figure 4.10 MIDI specified connector and pin assignments.

messages using other connectors and cables and due to the limited space on many PC adapters, many manufacturers use either a serial port or a joystick port to connect to MIDI instruments. A few MIDI instruments are actually equipped with an 8-pin mini DIN serial port, which makes it possible to connect those devices directly to some computers. However, the MMA (MIDI Manufacturers Association) does not currently approve the use of any other connectors for MIDI 1.0. Furthermore, although many Sound Card MIDI adapters are available, not all are designed according to the electrical standards defined by the MMA.

Data interface

The MIDI data stream is a unidirectional asynchronous bit stream transmitted at 31.25 Kbits/s. Note that many MIDI keyboard instruments include both the keyboard controller and the MIDI sound module functions within the same unit. In these units, there is an internal link between the keyboard and the sound module, which may be enabled or disabled by setting the 'local control' function of the instrument to on or off, respectively. When set to off, the instrument will not sound when it is played, but MIDI messages representing the performance will be transmitted in the normal way, allowing other sound modules to be played remotely.

The limitation imposed by the speed of the interface can cause problems, especially with the advent of more complex and real-time controls being made over instruments in the studio. For instance, it is not uncommon to perform all operations on a musical score within a computer running some form of sequencing software. The software may control all aspects of many sounds, including very slight variations in obscure real-time parameters. Some or all of the parameters of the sound may change constantly throughout a performance, and each slight change will be sent to the instrument via a new MIDI message. When this happens for a number of different parameters and musical instruments on the same MIDI network, then the number of messages quickly mounts up, and some messages may be lost as devices correct timing problems, or playback occurs in an untimely fashion. In the worst cases, careful routing should be considered so that data are selectively transmitted over different cables.

MIDI is particularly interesting in terms of transfer, since it is the first example from the audio industry to introduce any real addressing information into the bit structure of the data.

The physical interface is divided into 16 logical channels by the inclusion of a 4-bit channel number within the applicable MIDI message types. A MIDI enabled device, such as a musical instrument keyboard, can generally be set to transmit or receive on any one of the 16 MIDI channels. A MIDI sound source, or sound module, can be set to receive on specific MIDI Channels. In the system depicted in Figure 4.9, the sound module would have to be set to receive the channel that the keyboard controller is transmitting on in order to play sounds.

Although only 16 channels are available in the original MIDI specification, some sequencer software can support enhanced versions of the MIDI specification, allowing multiple networks and increased channel numbers, thus allowing instruments to be split onto physically separate networks. In such cases, each physical network will support 16 channels exactly as per the MIDI 1.0 specification. These enhancements are usually in the form of software routing programs that can assign a MIDI message output to one or other hardware MIDI physical interfaces designed to work with the software.

Addressing

The limitation for the number of channels comes from the 8-bit status information header of each frame, as described in the specification. The first 4 bits of any MIDI frame indicate the message type, and the second 4 bits indicate the channel number. Since only 4 bits are assigned for the channel number, the maximum number of channels that can be represented is 2^4 (i.e. 16).

Figure 4.11 illustrates the structure of MIDI messages, which can be up to 3 bytes in length, although not all messages require the frame to be this long.

Each device on the network will be assigned to a channel number, although this does not have to be unique for each device, allowing multiple devices to respond to any particular message.

Using Figure 4.9 as an example again, two devices are connected together in point to point fashion, using a single MIDI cable.

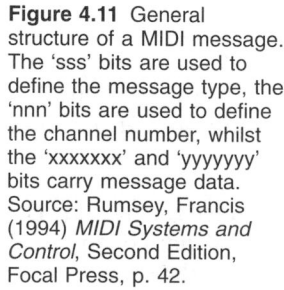

Figure 4.11 General structure of a MIDI message. The 'sss' bits are used to define the message type, the 'nnn' bits are used to define the channel number, whilst the 'xxxxxxx' and 'yyyyyyy' bits carry message data. Source: Rumsey, Francis (1994) *MIDI Systems and Control*, Second Edition, Focal Press, p. 42.

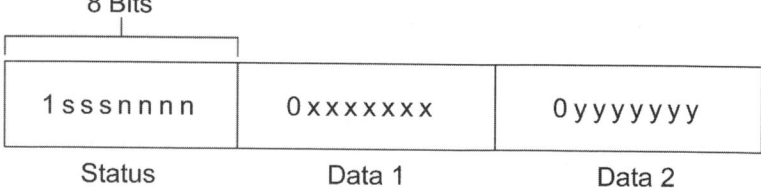

A performer plays instrument A, and this device is set to transmit information about the performance to MIDI channel 07. Provided that instrument B is set to receive on channel 07, then it will respond to the data being sent from instrument A, and will also sound. A maximum of seven devices can be attached in this way and if each device is set to receive to channel 07, each device will sound in response to the performance.

4.3.2 AES-24

AES-24 makes possible the control and monitoring of, via a digital data network, different audio devices from disparate manufacturers using a unified set of commands within a standard format (AES-24-1-1999, 4.1 Function – see Notes and further reading).

General description

AES-24 is a recent work undertaken by the Audio Engineering Society. It is an application layer protocol, as described by the title, although links to other layers cannot be avoided, and the relevant definition in context of the 7-layer model is 'With appropriate software, AES-24 commands are capable of being carried by most modern transport networks'. (AES-24-1-1999, 0.8 Transport networks – see Notes and further reading). This suggests that the full AES-24 committee work will cover aspects of the top three layers of the 7-layer model.

Standards and administration

The Audio Engineering Society has a structure of technical committees responsible for discussing different areas of audio development. One such council is titled the Technical Committee on Networking Audio Systems.

Each council may spawn a working party for any area that it feels is worthy of more investigation, and each such committee may decide that certain aspects of their scope are better left to other subcommittees, in order that best efforts are concentrated within a particular area, as intricacies unfold.

SC-10, formally known as the Audio Engineering Society Standards Committee (AESSC) SC-10 Subcommittee for Sound System Control, is the only control-related standards group within the AES. The group was the outgrowth of an earlier AESSC working group titled the WG-10 Working Group for Sound System Control, formed in the early 1990s.

In 1992, following the successful publication of AES15–1991, AES Recommended practice for sound-reinforcement systems –

Communications interface (PA-422), the SC-10 Subcommittee on Sound System Control saw the necessity to upgrade the standardization of sound system control to use the higher speed networks and efficient programming techniques then coming into use. SC-10 put forward a vision of a protocol that would be extensible and interoperable. Because the next available AES numbered standard was then AES24, the new protocol was dubbed AES-24 (Audio Engineering Society, Inc., 1999).

AES-24 is intended to produce four definitions, whose titles explain their roles in the overall development of the single protocol:

AES-24-1 AES standard for sound system control – Application protocol for controlling and monitoring audio devices via digital data networks – Part 1: Principles, formats, and basic procedures.

AES-24-2 Data types, constants, and class structure.

AES-24-3 Transport requirements.

AES-24-4 Internet protocol (IP) transport of AES-24.

Part 1 of the standard explains the concepts and defines a hierarchical structure, in the fashion of object orientation. The application of this structure is described in more detail in Part 2 of AES-24. Although the remaining two parts are not fully defined, Part 3 would appear to describe the necessary QoS and interfaces to the transport layer, and Part 4 to consider the use of IP as a network protocol. Unlike MIDI, no attempt is made to define a physical interface.

Object hierarchy

The protocol imagines devices consisting of objects. Each object will perform some control or monitoring function, such as increasing, decreasing or monitoring the output level of a particular audio channel. Some or all of the objects are presented to the network and those that are made available can be controlled by messages from the network, or can transmit messages onto the network.

Each object is an instance of a class, and objects contain methods, parameters, and events. Each object is identified by an address unique within the device, and each device will therefore require a unique address on the network (again unlike MIDI, but similar to digital data networks). Each object also contains data that defines its status. The objects presented to the transport network by a device define the logical transport interface of the device.

Considerations

Without strong support from manufacturers, the momentum for a common control protocol came from the consulting and systems integrator community, who stood to benefit the most from a standard.

Without a clear ability to profit from a standard, it was difficult to keep the attention of the very companies for whom it was intended. In the end, AES-24 languished, although some of the ideas presented within the committee structure have been adopted within other significant industry initiatives.

4.3.3 Advanced control network

In recent years, a parallel effort to AES-24 has evolved within the Entertainment Services and Technology Association (ESTA – see Notes and further reading). Their work on a common control method for entertainment systems is now headed in a similar technical direction to AES-24. In contrast to the SC-10 AES-24 effort, approximately 50 companies from the entertainment industry financially support the ESTA Technical Standards Program.

The advanced control network (ACN) is intended to provide the next generation standard for the distribution of data in lighting control networks. However, ACN is not limited to lighting with work undertaken for audio support for control and stage automation.

The advanced control network (ACN) protocol task group has the direct involvement of nine companies who provide engineering-level support for the development effort. Most notably, the ACN task group has asked for liaison support from the AES in their effort to include sound system control in their protocol.

Michael Karagosian (MKPE Consulting – see Notes and further reading) was appointed the chair of SC-10-02 and also worked with the ACN group within ETSA, creating a coupling between the standards-making groups, and allowing the work performed by either to cross over.

4.3.4 Conclusion to control data

MIDI deals with the physical layer and assumes a dedicated network, whilst AES-24 makes no presumptions about how the messages arrive. Extending this, ACN specifies IP as the interface of choice, allowing the protocol to exist on many different media and network types. This means that the format of messages and how they can be interpreted is concentrated upon, allowing IP to handle the network and transmission functionality.

MIDI was limited by the technology that was available and yet solved the whole interconnection problem for a limited set of instructions, whilst AES-24 and ACN build on modern techniques to create a more flexible set of instructions, leaving the lower layer technicalities alone.

In terms of Internet working, MIDI has an address space of only 4 bits, making the total possible number of addresses just 16. AES-24 and ACN assume that a modern data network can be capable of carrying the messages, and makes no comment on their transport.

4.4 Synchronization and timecode

Synchronization information is intended to allow several devices to operate with a common understanding of time. This is especially required in video applications, for instance where speech needs to match correctly with the video, so that when someone on the screen is talking, the words sound at the same time. Any delay beyond a human perceived tolerance will be immediately noticeable and result in a less than perfect experience for the viewer.

Synchronization is extremely important in the audio and visual fields, and especially so in post-production, which involves a combination of both fields. Several synch standards are in use within the audio industry, with several more related to the visual industries. Specification is through timecode, which allows the identification of a particular position in time, related to a start position used for locating a particular video frame in a recorded sequence for instance.

A timing signal can be generated in two main ways. The first method is to use a dedicated synchronization reference signal generator designed for the job. There are two main advantages to using such devices. The first is that they are often designed to send out a variety of different synchronization signals, each synchronized exactly to the other, since each signal is generated from the same timing. The second advantage is the supply of multiple ports on the unit with which to connect to other devices. In this way, the signal need not be looped through several devices in a single chain, and can be patched into different equipment with ease.

The second method for providing a clock signal is by assigning one of the items of equipment to become the master clock signal. Most items of equipment, such as DAT recorders and DAWs that have a need to be synchronized, already come supplied with an internal clock that can be used to send to other devices by setting the device to master and all the other devices to slave. This method does not usually include any kind of information about time, relative to an actual time, or any other fixed point (such as start of reel, as in the relevant sections).

This clock master arrangement commonly provides an immediate understanding of time between two devices, but does not generally provide a tool for positioning in time, and so the structure of the signal may be somewhat simplified compared to timecode proper, and is more correctly referred to as a clock signal.

Synchronization techniques take many forms and different mechanisms are used for different purposes. It is not the intention to describe all the intricacies of each technique, since this subject would fill a complete book on its own (Ratcliffe, 1996).

Instead, a selection of techniques will be examined in the context of shared access, addressing, and distribution.

As we have seen from previous chapters, it is necessary to synchronize two devices that are exchanging information in order that streams of binary information can be interpreted accurately over time. At this level, synchronization is required in order that two communicating devices understand where the boundary for each bit lies, so that consecutive bits can be identified and interpreted.

To achieve the identification of bit boundaries, techniques such as bi-phase mark encoding identifies the bit boundary with a transition in the bi-state, from high- to low-level signals for instance.

Such techniques become an integral part of the stream of information, and perform the specific requirement of synchronization between the two devices for the purpose of communication, but do not necessarily perform any synchronization with any other part of the process of audio or A/V production.

4.4.1 The Society for Motion Picture and Television Engineers

SMPTE is an international, technical society devoted to advancing the theory and application of motion-imaging technology including film, television, video, computer imaging, and telecommunications.

The Society was founded in 1916 as the Society of Motion Picture Engineers and the 'T' was added in 1950 to embrace the emerging television industry. The Society is recognized as a leader in the development of standards and authoritative, consensus-based, recommended practices and engineering guidelines.

The synchronization standard used throughout the audio and visual industry has the title SMPTE 12M–1995: for Television, Audio, and Film – Time and Control Code (available from http://209.29.37.166/stds/index.html email: smpte@smpte.org) and specifies a digital time and control code for use in television, film, and accompanying audio systems operating at 30, 25, and 24 frames per second.

Within the standard, time representation in a frame-based system is described within clauses 4, 5, and 6. To illustrate the problems of understanding time, clause 4 begins with a definition of NTSC time. NTSC is the American system of television, and NTSC time is defined against real time as 1 s NTSC time = 1.001 s real time. Whereas, in those sections that cover 25 and 24 frames per second systems, 1 s is defined as being exactly the time taken to scan 25 or 24 frames, respectively.

Clause 7 of the proposal describes the structure of the time address and control bits of the code, and sets guidelines for the storage of user data in the code. This consists of 16 4-bit groups, split into two halves of eight 4-bit groups. The first of these contains the timecode represented as hours, minutes, and seconds in such a way as to use 26 bits. The remaining 6 bits are used to determine the operational mode of the time and control code. Clause 8 specifies the modulation method and interface characteristics of a linear timecode source.

Clause 9 specifies the modulation method for inserting the code into the vertical interval of a television signal and clause 10 summarizes the relationship between the two forms of time and control code.

SMPTE reference signals include several standards such as: SMPTE 303M, SMPTE RP 154–1994, SMPTE RP 176–1993, SMPTE 274M–1995, SMPTE 295M–1997 and SMPTE 296M.

During 1998 discussions within the AES SC-06-02 working group on IEC61883-6 (mLAN – see IEEE 1394), the proposal noted that traditional SMPTE timecode may not have sufficient precision or accuracy to act as a timecode reference going forward. Furthermore, it was noted that adoption of a global time standard by SMPTE might fit well into further refinement of timecode standards. The SMPTE Workgroup: SMPTE Reference Signal

Workgroup, began work on this and proposals centred on using GPS time and navigational data (SC-06-02-C task group on Synchronisation in IEEE 1394 – see Notes and further reading).

It should be noted that different visual standards have different bit assignments. In the 625-line 50 Hz system used in Europe (PAL) and the 525-line 60 Hz system used in the United States (NTSC), bits 10, 11, 27, 43, 58, and 59 do not carry time or user data.

4.4.2 Longitudinal timecode (LTC)

When videotape recorders were initially developed, a voice-quality audio track was incorporated along with the high quality audio track, designed for the purposes of talkback. The playback rate of 30 frames per second, along with the bandwidth of the track, allowed a digital signal of 2400 bits/s to be recorded. This yields an 80-bit word, permitting 2^{80} combinations to be represented in each frame.

LTC requires 26 bits (2^{26}) to represent frame accurate time in hours, minutes, and seconds. The spare capacity becomes user-assignable bits, and these are grouped together in 4-bit words and are used for a variety of purposes, such as date, take or reel numbers.

When applied to video and audio recording, the code can be continuous and evenly spaced and runs the length (or longitude) of the tape, much like a system of evenly placed pulses. Alternatively, the code represents the time of day, which is the implementation used most often on multiple reel video production. In this case, information included within the code represents the time of day that the video sequence was shot. Each code word starts at the clock edge, immediately before the first bit (bit 0) with 80 bits per frame. The available data rate varies depending upon the implementation, so for instance 24 frames/s systems achieve a rate of 1920 bits/s, whilst 30 frames/s systems yield 2400 bits/s.

The second application for LTC is within television recorders, where different international standards have resulted in variations on the original.

The EBU and the Society for Motion Picture and Television Engineers (SMPTE – see Notes and further reading) standards have been incorporated into IEC standard 421:1986. This has been implemented within the EBU as standard N12:1994, in the UK as BS 6865:1987, and in the USA as SMPTE 12M:1995 (which also encompasses High Definition Television uses). The bit assignments for the entire 80-bit packet are shown in Figure 4.12.

Byte 1	Byte 2	Byte 3	Byte 4	Byte 5	Byte 6	Byte 7	Byte 8	9	10
Frames Units 1st Binary Group	Frames Tens 2nd Binary Group	Seconds Units 3rd Binary Group	Seconds Tens 4th Binary Group	Minutes Units 5th Binary Group	Minutes Tens 6th Binary Group	Hours Units 7th Binary Group	Hours Tens 8th Binary Group		

Synchronization ⌐

Figure 4.12 SMPTE LTC bit assignments within the 10-byte word. Eight of the bytes carry time and control data, and two carry synchronization and direct information. Source: Ratcliffe, John (1997) *Timecode, A Users Guide*, Second Edition, Focal Press, p. 32.

4.4.3 Vertical interval timecode (VITC)

One limitation of LTC that may not be immediately apparent is that it is difficult to read when the tape is stationary or being moved very slowly, such as may occur when trying to find an exact point within the tape in which to perform an edit.

The solution is to use those lines within the vertical scanning interval within the visual medium which are not used to carry test signals or other information to carry a new timecode. When the tape is stationary, the rotating head can read and therefore generate the timecode information.

As the active line period in 625/50 and 525/60 systems is around 52 µs, it is possible to incorporate a 90-bit code into one or more spare lines; however, the actual lines specified within the two systems is different. Data begins with two synchronizing pulses followed by 8 bytes of time and user information, which is in turn followed by 1 byte of cyclic redundancy check code as shown in Figure 4.13. The CRC word takes the place of the synchronization

Byte 1	Byte 2	Byte 3	Byte 4	Byte 5	Byte 6	Byte 7	Byte 8		
Frames Units 1st Binary Group	Frames Tens 2nd Binary Group	Seconds Units 3rd Binary Group	Seconds Tens 4th Binary Group	Minutes Units 5th Binary Group	Minutes Tens 6th Binary Group	Hours Units 7th Binary Group	Hours Tens 8th Binary Group		

Synchronization ⌐
Cyclic Redundancy Check Code

Figure 4.13 The VITC word comprises 8 data bytes containing time and control information, followed by a single byte for error detection. Each byte is preceded by two synchronizing bits. Source: Ratcliffe, John (1997) *Timecode, A Users Guide*, Second Edition, Focal Press, p. 42.

word used within LTC and although there is no automatic error correction, other techniques are employed to improve the immunity to errors.

4.4.4 AES11

Understandably, the AES also has some recommendations regarding synchronization of digital audio signals, and these are to be found in the document AES11–1991. The document describes synchronization in terms of sample frequency and phase synchronization and recommends that all machines should be synchronized to a reference signal taking the form of a two-channel interface signal of a stable frequency and within defined tolerances. The paper also recommends that each machine has a separate input for the synchronization signal.

Multiple signals are considered to be synchronous when they have the same sampling rate, although small differences are accounted for by allowing phase errors to exist between the reference clock and the digital audio data. This allows for effects such as propagation delay in the cables and so on.

The frame boundaries at the start and stop of the frame of the input signal must be within ±25% of the reference signals frame boundary and the output should be within ±5%.

Two grades of signal are specified, grade 1 and grade 2. Grade 1 has a specified long-term frequency tolerance of ±1 ppm and is intended for larger facilities, which may run several studios from a single reference signal. Grade 2 is intended for single studios, where a greater need for accuracy is not required, and specifies ±10 ppm.

4.4.5 MIDI and timecode

As we have seen, there are two types of MIDI message: system and channel. System messages start with a 4-bit system message code, followed by a message type code. A system message has its header set to 1111, and 16 types of system message are available (although not all are implemented). MIDI clock signal is transmitted relative to the rate of the music being played through the MIDI interface and is defined as 24 clock periods per quarter note (ppqn – crotchet). These messages are best thought of as a metronome, as they contain no time-related information. The start message causes a sequencer to start playing its sequence from the beginning, and the stop message stops the sequencer playing. The continue message is used to tell the sequencer to start playing at whatever position in the sequence the pointer has reached. The pointer can be positioned anywhere within the sequence by counting the intervals from the start of the sequence.

Synchronization between externally generated clock signals and MIDI can be achieved, although this is sometimes difficult because of the tempo-related clock signal used within MIDI.

It is often a requirement for equipment using MIDI and IEC timecodes (such as longitudinal timecode and vertical interval timecode covered in more detail in the following sections) to be interfaced, especially in post-production environments. To accommodate this, a system of representing real-time within MIDI has evolved called MIDI timecode (MTC).

MIDI timecode

MTC is specific to MIDI enabled equipment and is designed to be transmitted over the MIDI interface. Most modern MIDI interfaces can convert longitudinal timecode to MTC for the purpose of synchronizing a computer sequencer to a tape or other recording machine.

In order for MTC to be accepted by MIDI machines, it must take the same form as other MIDI messages. Therefore, it must have a status byte and data bytes. There are two main types of message sent over MTC. The first and most common is the running timecode and is known as a quarter-frame message. This can be likened to sending seconds over the MIDI interface.

The second type of message is a one-off information message such as might be sent when rewinding a tape, to indicate the time position that has been reached during the rewinding process. This is sent as a universal system exclusive (sysex) message. Each message type is identified by the header byte.

MTC quarter-frame messages

The quarter frame message is preceded by a Systems Common Header (hF1) and is used to send out real-time data and so is regular over time.

Longitudinal timecode and vertical interlude timecode assigns two binary groups each to hours, minutes, and seconds. This is too much information to fit into a MIDI frame, and so the information is split into eight different frames. In simplified terms, the frame is made up as illustrated in Figure 4.14.

The first bit is a zero, and the remaining 7 bits of the word represent whether the message contains frames, seconds, minutes

Figure 4.14 General format of the quarter-frame MTC message. Source: Ratcliffe, John (1997) *Timecode, A Users Guide*, Second Edition, Focal Press.

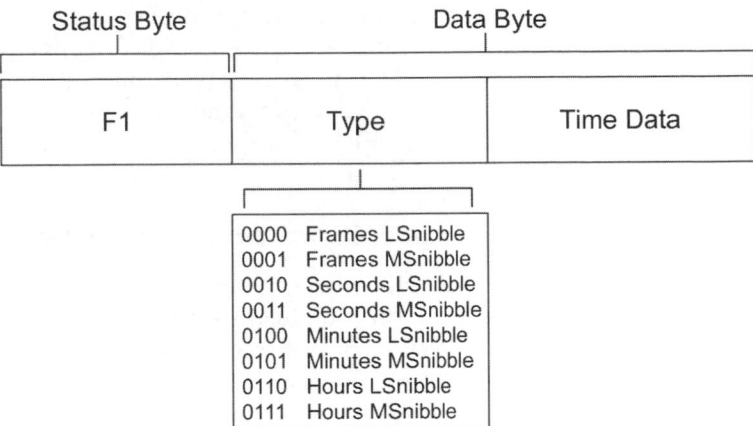

Status Byte	Data Byte	
F1	Type	Time Data

```
0000   Frames LSnibble
0001   Frames MSnibble
0010   Seconds LSnibble
0011   Seconds MSnibble
0100   Minutes LSnibble
0101   Minutes MSnibble
0110   Hours LSnibble
0111   Hours MSnibble
```

or hours. The next 4 bits represent the actual decimal value of that particular time division. To reassemble the data from the eight quarter-frame messages into an understandable time value, the messages are paired up in the receiving device to form 8-bit bytes.

Full-frame messages

The system exclusive header (F0h) precedes the full-frame message. Full-frame messages are generally used in situations where it is impractical to send the quarter-frame messages. An example of this would be during a fast rewind of a tape machine, where too much data would be sent over the slow link if a message was sent for every time position in the sequence which the pointer covered during the rewind process. In these instances, a one-time update of the current position of the tape (relative to time) is sent at representative intervals in a full sized frame of 10 bytes.

There are three types of full frame message. These carry real-time data, binary user group data associated with the IEC code, and finally set-up messages.

Full-frame messages are a minimum of 10 bytes, and are prioritized and so may be sent out in the middle of a message stream.

Synchronization with other sources

The problem with synchronizing MIDI to external synchronization signal sources is in converting tempo-related messages into real-time-related messages.

The two main approaches are to use a synchronizer with an integral timecode to MTC converter, or to use a dedicated synchronizer to convert timecode into MIDI clock or MTC, thus enabling the operation of a sequencer to control the remainder of the MIDI set-up.

4.4.6 Conclusion to synchronization and timecode

Unsurprisingly, none of the standard mechanisms for synchronization contain any form of addressing, since synchronization is a matter of distributing information about time. Therefore, distributing a synchronization signal is generally a matter of checking with a source reference, in the same way that weights and other standard units of measurement (including time) all relate to a source for reference. As a system of measurement, these sources are considered infallible and not subject to change, and all other references to the base measurement must be made against the source reference. All other references to that unit of measurement (such as rulers, weights, and clocks) can eventually be traced back to the source reference.

This can also be done with measurements of time, although this is obviously impractical when many electrical devices are distributed over a great distance. Instead, all the devices swap and compare their understanding of time, and negotiate amongst themselves to establish a common understanding. This is not as complex as it sounds and the method is frequently employed in both the computer industry and the audio industry.

The important concept is that the digital information is discrete, which in this case means that the process for encoding the analogue information is clocked against an assumed time. If the information is then decoded using a different measurement of time, the mechanism will not necessarily error (unless programmed to do so), but will instead assume that the time is the same. This provokes thoughts of playing back a record at a different speed, but the intention is to illustrate that if the two clocks are a tiny fraction of a second out of time with each other, then the result will be a less than perfect time relation during playback. In multi-track and A/V situations, this may result in audio tracks that are out of synchronization with the visual experience. However, if two machines are forced to compare clock signals with each other, then a compromise between the two can be reached. In audio production, one machine is generally assigned as the master or source reference and the other machines must follow.

In computer networks, another method used for achieving an understanding of time results in devices voting on what the time is. In this case, certain rules arbitrate the voting system, meaning that not all the attached devices get to vote. For instance in some networks, segments of the network that are partitioned from the rest of the network by a slow speed link will have their votes discarded, or may not be entitled to vote at all (Novell Netware 4+ Operating System). The agreed time will be equal to the average of the votes. If only two devices were attached to the network, then the time would be half the difference in time between the two machines.

Voting schemes are used in many instances within computer networks to reach agreements on various aspects of network operation. In token ring networks, for instance, the first device to identify that the token is not being transmitted will send out a request to vote on which device should become the token master.

4.5 Conclusion to audio communications

The international standardization of data structures and encoding methods (such as PCM) in digital audio data is useful since it represents a solid final format that is readily understood and utilized. The advantage is that when all the transport information and encoding is discounted, the audio data remain in their internationally agreed digital form. This is useful since the data can be surrounded and encoded in a new way to be transmitted from one device to another using a different interface. A simple form of this processing would be in a patchbay, designed to strip one protocol and add another, thus translating from one interface to another. This functionality is similar to network bridges and routers, which allow data to be transmitted over different kinds of medium.

This has been used to some effect with matrix routers using the MADI recommendations. Although this standard does not specify any addressing information within the frame structure, the identity of an audio channel can be ascertained by its position within the super-frame, and so routing can occur, since the source of the channel can be identified in this way.

Notes and further reading

AES-24-1–1999. 4.1 Function, page 16. 1999-03-02-1 print. Audio Engineering Society, Inc., 60 East 42nd Street, New York 10165, USA. http://www.aes.org.

AES-24-1–1999. 0.8 Transport networks, page 6. 1999–03-02-1 print Audio Engineering Society, Inc., 60 East 42nd Street, New York 10165, USA. http://www.aes.org.

Audio Engineering Society, Inc. (1999) *Trial-Use Release of Proposed Sound System Control Codes*.

Audio Engineering Society, Inc. International Headquarters, 60 East 42nd Street, Room 2520, New York 10165–2520, USA. AES recommended practice for digital audio engineering – serial transmission format for linearly represented digital audio data.

ESTA Administrative Office. 875 Sixth Avenue, Suite 2302, New York 10001, USA. http://www.etsa.org.

Held, Gilbert (1998). *Voiceover Data Networks*. McGraw Hill Text.

IEC60268-12 (1987–03). Sound system equipment. Part 12: Application of connectors for broadcast and similar use. Central Office of the IEC. 3, rue de Varembé, PO Box 131, CH-1211, Geneva 20, Switzerland. http://www.iec.ch.

IEC 60958-1 (1999–12) Digital audio interface – Part 1. General Central Office of the IEC, 3, rue de Varembé, PO Box 131, CH-1211 Geneva 20, Switzerland.

MIDI Manufacturers Association Incorporated. PO Box 3173, La Habra, CA 90632-3173. http://www.midi.org.

MKPE Consulting. 23679 Calabasas Road 519, Calabasas, CA 91302-1502, USA.

Ratcliffe, John (1996) *Timecode, A Users Guide*. 2nd edition. Focal Press.

SC-06-02-C task group on Synchronisation in IEEE 1394. *Report of Synchronisation Task Group to SC-06-02 regarding project AES-X60*. 9 September 1998.

SMPTE. 595 West Hartsdale Avenue, White Plains, NY 10607, USA. http://www.smpte.org.

5 Manufacturers' interfaces

There are many manufacturer interfaces in use throughout the audio industry. For example, TDIF from Tascam and SDIF-2 from Sony are two more that are not covered in great detail in this book (as the technology is out of context, or the concepts are already thoroughly covered within).

Manufacturer interfaces quite often start life as a research project for a commercial organization and so are subject to trademarks, patents, and copyright. A manufacturer may choose to open the work for wider development, or keep the technology proprietary. Those that are opened may request administration by a standards body, and so become adopted, as with IEEE 1394.

Other mechanisms are patented by the investing manufacturer and implemented as part of a device. The sponsoring manufacturer normally expects the interface to connect to other equipment, also supplied with the proprietary interface.

The third alternative to manufacturers, who wish to promote a proprietary technology, is to freely publish extensive descriptions in the same way that a standard might be published, encouraging other vendors to adopt the solution and develop equipment that supports it. This is similar to the situation that developed with FireWire from Apple Computers, except that the technology is not necessarily adopted by the standards-making bodies and so relies on the generosity of the development agency. Successful examples are those that are adopted within the vertical industry and can be described as de facto standards.

TDIF is an example of a de facto standard and, like ADAT, this also supports eight channels of digital audio, transferring over a cable connecting to a 25-way D-type connector. Although an RS232 connector is used to physically connect devices together, the transfer is not serial in nature, as might be expected from a physical connection so commonly used for serial communications on desktop computers.

5.1 ADAT

Alesis digital audio tape (ADAT) provides a mechanism for the transportation of digital audio data from Alesis ADAT recorders to another device.

5.1.1 General description

The patent describes an interface in which the sampling rate and number of audio channels can be discovered by the receiving device. This is achieved by ensuring that a pattern exists within the data stream, to which the receiving device can lock on to. The transmitted data must therefore contain the synchronization pattern in order for the receiving device to correctly interpret the incoming data stream.

5.1.2 Standards and administration

The Alesis Corporation (see Notes and further reading) maintains responsibility for the interface, since ADAT remains a proprietary technology. The details of ADAT are recorded as a patent and can be inspected at the United States Patent Office website (United States Patent: 5 297 181 – see Notes and further reading).

5.1.3 Encoding

The data encoding mechanism uses a variation on the NRZ bit-encoding mechanism called NRZI, which is used to supply a period of zero volts (and no transitions) within the data stream.

Audio channels are encoded using the PCM mechanism and the transmission mechanism can automatically adjust itself to whatever sampling frequency is used, by controlling the other parameters in use, such as the number of bits in a word.

Although the patent is flexible in its application, the ADAT interface will specifically interpret resolutions up to 24-bit word length and 48 kHz sampling frequency, with scope to allow a wider range of resolutions to be accommodated.

Synchronization Pattern

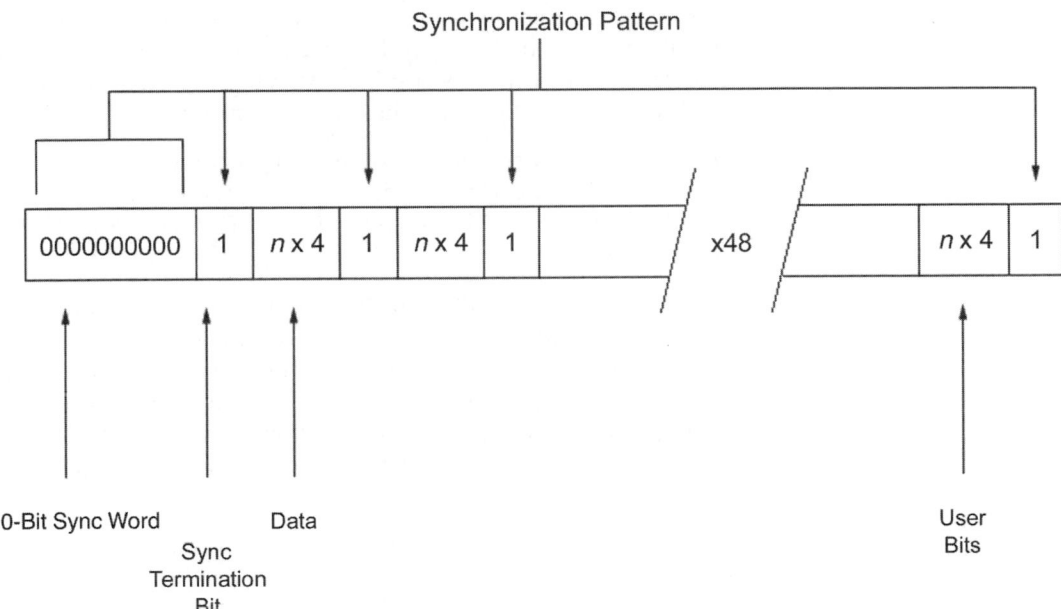

| 0000000000 | 1 | n x 4 | 1 | n x 4 | 1 | | x48 | | n x 4 | 1 |

10-Bit Sync Word   Data

Sync
Termination
Bit

User
Bits

Figure 5.1 Alesis ADAT
Lightpipe frame, showing
synchronization pattern within
the frame.

5.1.4 Framing

Data frames of 256 bits, with an 8-bit clock counter, simplify the frequency recognition stage.

Illustrated in Figure 5.1, the frames are made up of eight groups of 4 bits, making a total of 192 bits of data. Into this frame is also inserted the synchronization pattern as well as user data to make up the 256 bits for the frame.

5.1.5 Frequency detection

Frequency detection in this context means the detection of the sampling frequency by the communications interface. This is made possible because the sampling frequency is directly related to the data rate. As a result, the pattern to which the interface locks is contained within a frame.

The synchronization pattern is made up of a signature period of no signal (or transitions) at the leading edge of each frame as well as the regular placing of 1s within the rest of the frame. The period of no voltage is defined as ten consecutive 0s terminated with a binary 1. The regular placing of 1s within the frame is achieved by placement after each group of 4 data bits. The frame is also terminated with a 1, so that the period of no voltage (known as the 10-bit sync word) is encapsulated by transitions and can be timed.

The receiving device locks on to the incoming signal without any prior knowledge of the sampling frequency, by using a three-stage process shown in Figure 5.2 involving the identification of the synchronization pattern and then the frame.

Figure 5.2a shows the first, course adjustment stage. This stage involves finding the sync word, measuring its transmission time and adjusting the oscillation period of the VCO to meet some fixed parameters.

The process dictates that if more than 11 consecutive zeros are counted, the frequency of the VCO within the receiving device is decreased. If fewer than eight consecutive zeros are counted, the frequency of the VCO in the receiving device is increased. Once the frequency of the VCO is within these limits, control of the VCO is passed to stage 2, which is known as the fine adjustment and is illustrated in Figure 5.2b.

Since a regular period has already been identified in stage one, the fine adjustment involves counting 256 oscillations for every occurrence of the sync word, thereby identifying the individual bits. To do this, a counter counts VCO oscillations between consecutive sync words. If more than 257 clocks per sync word are counted, the frequency of the VCO is decreased. If fewer than 255 clocks are counted per sync word, the frequency of the VCO

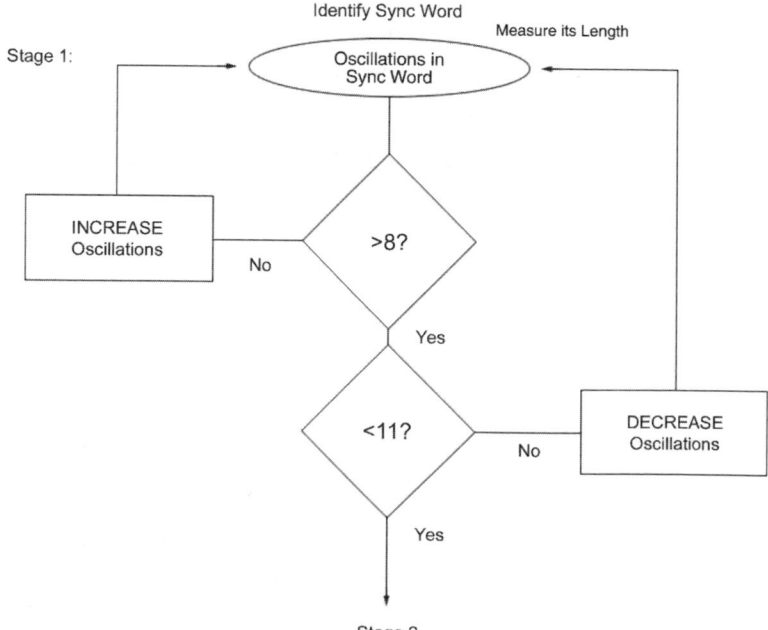

Figure 5.2 ADAT frequency detection. Each stage represents a step closer to identifying the frequency. (a) Stage 1 identifies a frame from the frequency of a recurring pattern.

Figure 5.2 (b) Stage 2 identifies the correct amount of bits within the frame.

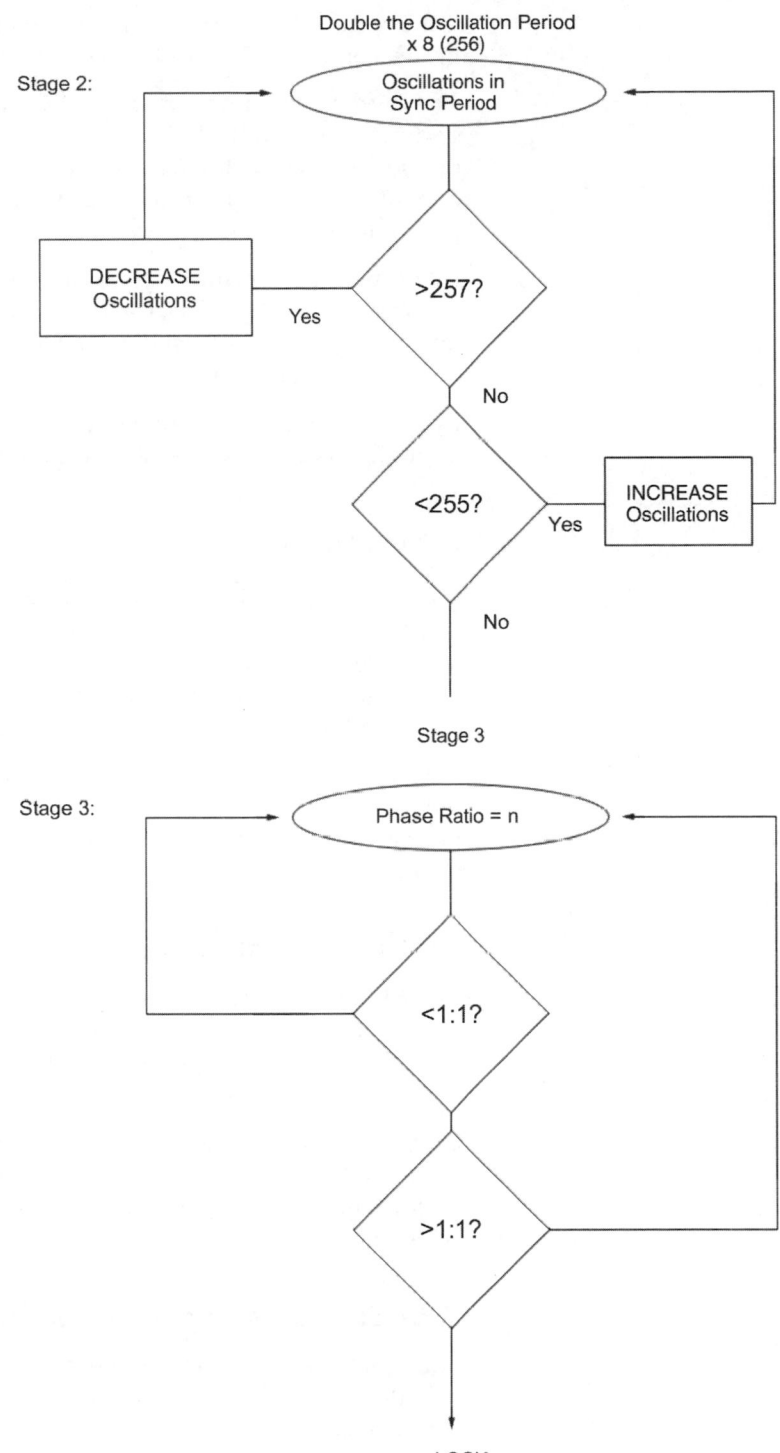

Figure 5.2 (c) Stage 3 alters the phase to achieve a successful frequency detection.

is increased. When the frequency of the VCO is within these limits, control of the VCO is passed to stage 3.

The third stage, shown in Figure 5.2c, controls the phase of each of the receiving unit's VCO. A phase detector compares the phase of the VCO clock to that of the data stream. If the phase of oscillation within the receiving device is ahead or behind the phase of the data stream, then the VCO is adjusted accordingly.

In essence, the synchronization takes place by the recognition of a period of the same electrical or optical state. In order to do this, long strings of zeros must be avoided, and so data are grouped in fours, each group being appended with a fifth bit, set to 1.

By careful application of logic, the ADAT interface achieves approximately 75% efficiency of the interface.

5.2 CobraNet

Information in this section is reproduced with the kind permission of Peak Audio (copyright 2000 Peak Audio Inc. – see Notes and further reading). Peak Audio devised a scheme to move digital audio data around on a physical infrastructure complying with the 100 Mbits/s Ethernet recommendation.

The idea of moving data with such deterministic QoS requirements as real-time audio on Ethernet technology warrants further investigation, as Ethernet was dismissed as unusable in earlier chapters.

5.2.1 General description

In addition to carrying audio and the associated sample clock, control data can also be carried over the network. Furthermore, CobraNet allows control and monitoring schemes from different vendors to co-exist on the same network infrastructure.

On repeater hub networks, CobraNet orchestrates data transmissions and eliminates collisions. Without collisions, real-time performance and higher utilization of an Ethernet network is achieved.

5.2.2 Standards and administration

CobraNet is trademarked by Peak Audio and is proprietary and patented within the US Patents Office. CobraNet is offered as a licensed technology for the transport of multi-channel audio and control data, and several companies have already subscribed.

Peak Audio has set up the CobraNet Manufacturers Consortium (CobraMC) through which the CobraNet standard is maintained. Membership is restricted to representatives of CobraNet licensed manufacturers. CobraMC conducts business through the Internet and meets twice yearly at the NSCA and AES conventions.

5.2.3 Physical

At the Physical layer, CobraNet networks adopt the 100 Mbits/s Ethernet cabling standard and are capable of delivering audio in blocks of eight channels at 16, 20 and 24 bit, 48 KHz, making the capacity of a 100 MB/s network equal to 64 audio channels.

The inability of Ethernet to deliver a deterministic QoS has been documented elsewhere. Consider also the burden imposed by upper layer error checking, such as those described for the TCP/IP protocol. The CobraNet solution is to prevent the random function from being invoked by imposing a new set of rules before the Ethernet CSMA/CD algorithm.

5.2.4 Arbitration

The additional arbitration layer for the CSMA/CD mechanism is known as the O-Persistent layer, and is positioned as shown in Figure 5.3. Devices on the network that do not implement the O-Persistent rules cannot be used on the same network, since

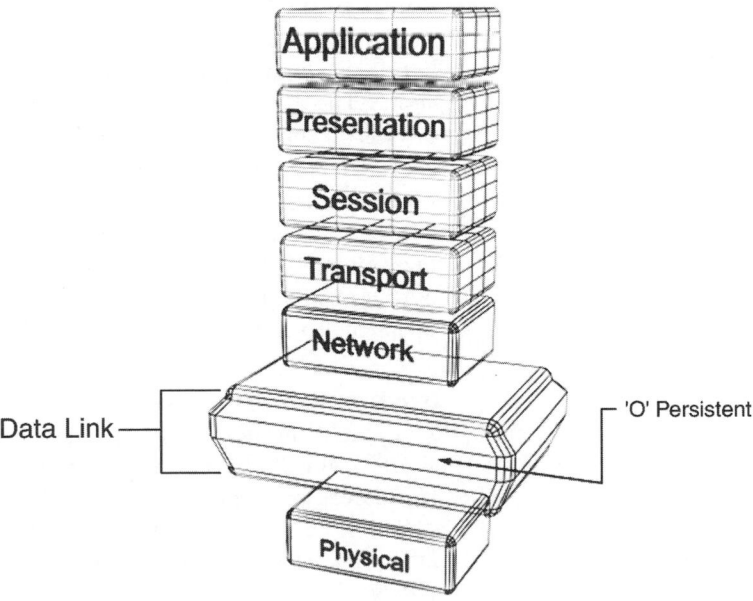

Figure 5.3 Peak audio CobraNet O-Persistent layer.

these will cause collisions and once again result in invocation of the undetermined waiting time or garbled messages at the very least.

Transmissions are pre-screened by the new sub-layer and the upper layers are discarded and replaced with new network layer solutions.

5.2.5 Switched networks

Although Peak Audio's initial implementation of CobraNet required the use of the O-Persistent layer in order to arbitrate, more recently it has been updated to support switched networks.

In switched networks, the effort required to arbitrate the network is not required, and coupled with the advantages of this kind of network, which are explored more thoroughly in Chapters 6 and 7, make for a considerable improvement in the efficiency and ability of the network.

5.2.6 Packet types

CobraNet uses three basic packet types. All packets are identified with a unique protocol identifier (0x8819) assigned to Peak Audio.

The beat packet

The first of the packet types is the beat packet, which contains network operating parameters, clock, and transmission permissions. The beat packet is transmitted from a single CobraNet device on the network and indicates the start of the isochronous cycle. The beat packet carries the clock for the entire network and so is sensitive to delivery delay variation. Failure to meet the delay variation specification may cause devices to loose synchronization with the network. The beat packet is only around 100 bytes on a typical network, although this increases depending upon the number of active channels.

Isochronous data packet

The isochronous data packet can be multicast or unicast (to a single station) and contains audio information. A small amount of buffering is performed in CobraNet devices so that late delivery of data packets is acceptable. In order to increase efficiency, data packets remain relatively large at around 1 Kbyte.

The reservation packet

CobraNet devices transmit the reservation packet once per second. These packets are crucial to the audio transmission and three dropped reservation packets in a row will cause an interruption in audio. Dropped data packets affect only the associated channel, whilst dropped beat packets affect the entire network.

Responsibility for the properties of each audio channel are entrusted to the upper layers, so that each channel in a frame can have a different bit resolution and sample rate. Data frames can contain data for up to eight audio channels. The size of the data packet varies depending upon the payload.

By ensuring that no collisions take place, or by using switched mechanisms, the delivery of data frames becomes accurate enough to match clock signals with audio frames to create isochronous streams from the remaining Ethernet algorithm.

The collection of the data into frames is a buffering activity, inevitably resulting in some delay. The delay is in the order $5\frac{1}{3}$ m/s (256 sample periods) on most installations and is constant from one end of the network to the other with A/D and D/A processes adding further delay.

5.2.7 Network design

By using switched technology to transfer audio information, thorough consideration needs to be given to network design and implementation, especially in large networks. Good documentation is available at the company's website (www.peakaudio.com).

5.3 Soundweb

Information in this section is reproduced with the kind permission of David Karlin of BSS Audio (see Notes and further reading), as Soundweb is a proprietary audio network technology from BSS Audio, which is most often used in audio installations such as convention centres and suchlike.

The technology covers the control of up to eight digital audio channels at 24-bit, 48 kHz resolution, over a physical category 5 cable. Each cable length can be up to 300 m in length and devices can be connected together directly or by using a hub. Each device on the network must use the Soundweb network access mechanism.

The audio processing devices are loaded into digital signal processors (DSP) contained within BSS Soundweb devices. These devices may be distributed around the network, although each parameter of each device can be controlled and monitored from a single PC connected to the network.

5.3.1 General description

The network specifies a bandwidth of approximately 12 MB/s, where three quarters of this is used to transport eight channels of 24-bit, 48 kHz audio and the remaining bandwidth is used for control data.

Real-time control of audio processing block parameters is achieved for any network configuration by designing a control interface (either in software or hardware) and the messages sent from such a tactile interface are encoded into the audio stream.

5.3.2 Physical

Media

Soundweb is designed to use category 5 cable for transmission, and cable lengths up to 300 m are supported. Standard RJ45 network cables, wired in the normal way (see Chapter 1), are used for network in and out connectors provided on each unit.

Extensions were added to the Soundweb specification in order to support fibre transport media, allowing the network to be extended to 2 km for a single run.

Topology

Soundweb topology is essentially similar to token ring. In cable terms, the ring is implied as the return path does not require a separate cable, but is cabled instead through separate wire pair inside the standard category 5 arrangement.

When connecting a fibre media adapter (9014 Fibre Interface, BSS Audio), the ring out and ring in are separate fibres.

Arbitration

One device will be the clock master for the ring, performing management of the token. The ringmaster is negotiated randomly on start-up in order to minimize problems caused by a cable failure, so that a new master can be easily negotiated in the new ring sections. The ringmaster sends tokens around the network allowing the clock slaves in each device to start operation.

Data link

The data separator receives the incoming data stream, and understands the frame format from within the state changes on the cable. The function performs the metamorphosis from physical media to data, and makes arbitrary decisions based on indicators held within the frame.

Encoding

Data are encoded onto a link using the Manchester encoding technique which describes a 1 as consisting of a half bit period in the low state followed by a half bit period in the high date; a 0 consists of a half bit period in the high state followed by a half bit period in the low state. Essentially, Manchester encoding specifies a state change for each bit period.

Synchronization

A deliberately illegal pattern that breaks the Manchester rules of encoding is used at the start of each frame. This is used to lock onto the signal for synchronization purposes. The pattern consists of the sequence HL LL HH, in which two consecutive bit periods do not contain state changes. Furthermore, the single change at the bit boundary allows better focus for the synchronization function. The illegality of the code allows the receiving device to recognize the pattern and so recover from a situation where it has locked onto a phase that is up to half a bit period out from the incoming data stream. This allows an arbitrary understanding of time whereby the receiving device is responsible for adjustment.

Framing

Data are transmitted as a series of 256 bit frames, each containing one sample. All quantities are sent LSB first, apart from audio data, which are sent MSB first.

When all eight channels are in use, the frame format is as shown in Figure 5.4.

Figure 5.4 BSS audio Soundweb frame structure.

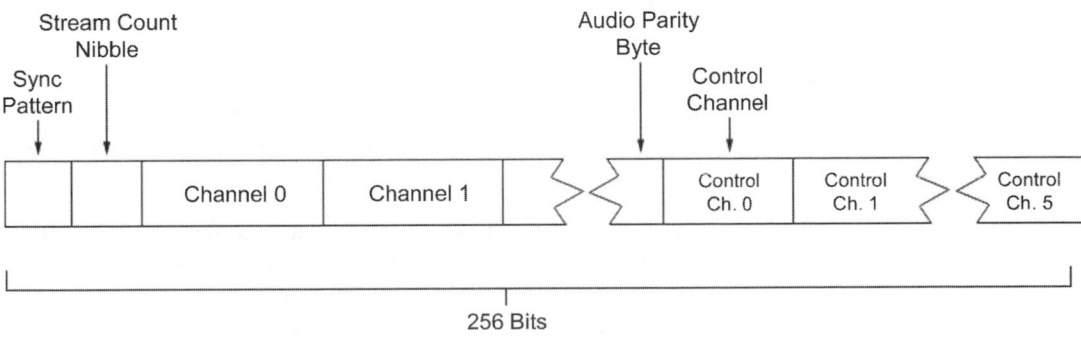

The sync pattern located at the start of the transmission uses exactly 4 bits to allow the receiving device to lock to the data stream.

The stream count nibble indicates the number of valid stream channels. After the stream data have been sent, 1 byte of parity bits is sent. The remaining bytes in the frame are used for control. When all eight channels of audio are in use, 6 bytes per frame are available for control. With no audio present, 31 bytes per frame are available.

Control protocol

Each audio sample transfers between 6 and 30 control bytes, depending on the number of active audio channels in the sample within the frame structure. The first byte in the frame is a flag field, indicating the nature of the remaining bytes in the frame.

Frames are grouped together into packets, which contain flag byte, address, byte counter, payload, padding and CRC.

Error handling

The CRC is the last 2 bytes in the last frame of a packet and contains an even parity bit for the audio data in each channel. The CRC is calculated for the whole packet, excluding the flag and address bytes. Each bit corresponds to one of the streams of audio data.

The byte count is treated as part of the data for the purposes of reception. The padding bytes are generated automatically. The packet size is limited by the amount of RAM allocated to transmit and receive areas of shared RAM, which are themselves limited to a total of 8 kilobytes by the specified address space.

5.3.3 Network

Addressing

Addressing on the network is by simple means of a counter. The behaviour of the counter is specified as decreasing by 1 at each node, in a reverse arrangement to hop counts and represents a cost.

When the counter reaches 0, the packet has arrived at its final destination. When a packet is broadcast onto the network, the counter is set so that the packet expires when it has gone out to all the nodes on the ring and is returned to the sender.

Arbitration

When no data are being sent on the network, the ringmaster sends out FRAME_EMPTY packets, which are forwarded round the ring until they arrive back at the ringmaster, where they expire.

If a device other than the ringmaster wishes to send data, it converts the FRAME_EMPTY to a FRAME_VALID, and appends data according to the frame rules. When all the data have been sent, the next control frame is set to FRAME_EMPTY.

The data are sent back round until they reach the ringmaster. If the destination has been reached, the ringmaster sends out a fresh FRAME_EMPTY packet. If not, the ringmaster forwards the data. If the ringmaster identifies the data next time around, the packet is expired and a new FRAME_EMPTY message is sent.

When the master wishes to send data, it waits for an expired FRAME_VALID frame, or for a FRAME_EMPTY that has come back round the ring. On receipt of either of these, the frame does not need to be forwarded and data can be sent out.

5.4 MediaLink

MediaLink is a Fairlight trademark, which also uses standard Ethernet 100BaseT network components. This time the whole protocol is rewritten to achieve 65% of network efficiency, capable of 80 channels of 48 kHz digital audio (*MediaLink Product Overview* – see Notes and further reading).

MediaLink uses a Windows NT operating system allowing it to interface in a gateway arrangement. It is not possible for either MediaLink or CobraNet to support other protocols on the same cable.

5.5 GMICS

Global musical instrument communication system (GMICS – usefully pronounced 'gimmicks') is an initiative to produce a digital connectivity standard aimed at live performance and is most closely described as 'performance oriented' in Chapter 2.

The goal of the specification is 'to enable musical instruments and their supporting devices such as amplifiers, mixers, and effects boxes from different vendors to digitally interoperate in an open-architecture infrastructure'.

5.5.1 General description

The GMICS engineering specification describes a system utilizing the physical media standards of 100Base-TX (from the 100 Mbits/s Ethernet specifications), the only differences being within the original definition, meaning that topology and media remain identical. The specification also defines the frames and network access mechanisms. The system works by using the principle of data moving asymmetrically (discussed in detail in Chapter 7). That is, even considering duplex data traffic, audio data are moving away from their point of generation and towards a new endpoint (hard disk, loudspeakers, etc.).

Multiple instruments may generate data toward a hub or mixer, which uses a TDM technique to stream data onward. The system connects together in the manner of recording or reinforcement systems, such as sound reinforcement for stage performances or within recording studios. Devices may be daisy chained in the same manner as guitar foot pedals, allowing downstream processing.

5.5.2 Standards and administration

GMICS is a trademark of Gibson Guitar Corporation. GMICS is promoted as a freely available standard, accepting public comment, with development encouraged. The specification is available for free download on the Internet (http://www.gmics.org; Jusszkiewicz *et al.*, 2000).

5.5.3 Physical

Multiple physical interfaces can be used with GMICS, based as it is on the 100 Mbits/s Ethernet physical layer. The 100 Mbit GMICS Link is called G100TX, which refers to the interconnecting cables and data-transport mechanism. The interface is defined as transportable across category 5 cable using RJ45 connectors. Further interfacing media can use the gigabit Ethernet standard for high speed multi-link 'backbones', also mentioned within the engineering specification.

Topology

As mentioned, GMICS assumes an asymmetrical flow of data, at least in terms of connections. Two different connection types are called Type A and Type B. To metaphor the analogue domain, these are connected in the same fashion as audio-in and audio-out, although the function is duplex data and not unidirectional analogue audio.

Synchronization and timing

Attention is paid to timing-related issues such as jitter and latency, going on to describe the quality of service performance expected from conforming devices.

The network synchronization is defined in section '2.5 Master Timing Control' of the engineering specification. This describes a system timing master (STM) as a non-instrumental device, operating downstream from the source.

System timing master

The selection of STM is based on system hierarchy, assuming a tree-like structure, whereby by the root node becomes the master.

The STM provides the sample clock, and enumerates the addresses of all other devices.

Enumeration

The address field is 16 bits wide, limiting the number of addresses to 65356. A default start-up address of 0xFFFC is specified, to which all devices must respond, so that communication can occur whilst the address is assigned.

Upon receiving its address, the device returns an address offset return. This acts as an acknowledgement, and accounts for use of hubs, to which more than one device may be attached. Each device in the chain performs the same actions and in turn each device waits for the address offset return message before returning its own. This allows flexible connection regimes as illustrated in Figure 5.5.

Recovery

Keeping in mind the conditions under which the network is expected to operate, the specification is designed to allow for devices to attach to, or become detached from, the network at any time.

If a new device is attached to the network following the network initialization (the powering up of the first devices when the STM is established), the STM assigns the next number in the sequence. If devices are removed from the network, the addresses within the STM are not reassigned.

5.5.4 Data link

Within GMICS, the packet contains a header and the data payload. The header information contains information regarding the status of the network, which allows for arbitration over the network.

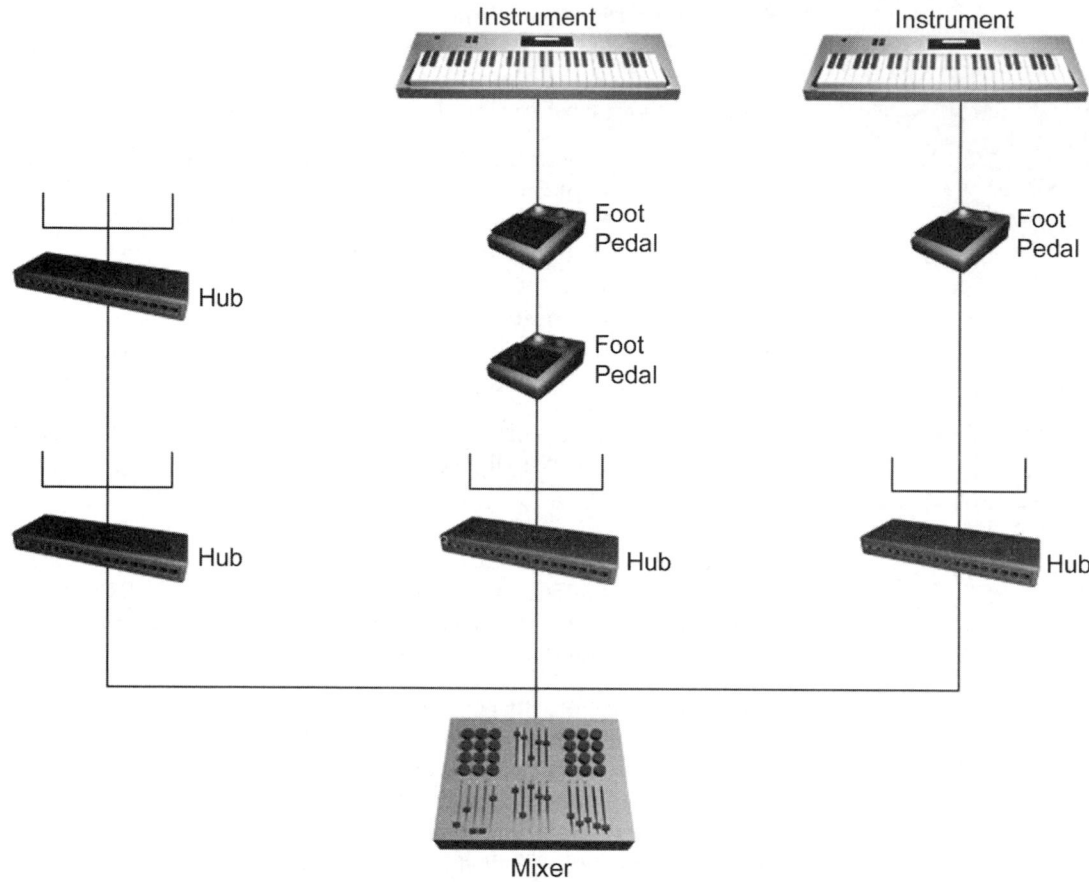

Figure 5.5 Connection topologies within GMICS specification.

Frames

GMICS frames are fixed at 27 32-bit fields in length

The header format is shown in Figure 5.6, and begins with the preamble and start of frame sequence specified in the IEEE 802.3 specification associated with CSMA/CD (see Ethernet). The header contains other information, including bus control indicators, packet type flags, a counter, and sample rate.

Bus control

Bus control is performed using the clear to send (CTS) and message in progress (MIP). The principle is that a device requesting to send a message sets the CTS flag to indicate to other devices that a new message is now queuing. Any message that is being sent has priority and will continue until its completion. No new messages can be queued until the CTS flag is reset. The CTS flag is reset by the device that set the flag in the first place, once

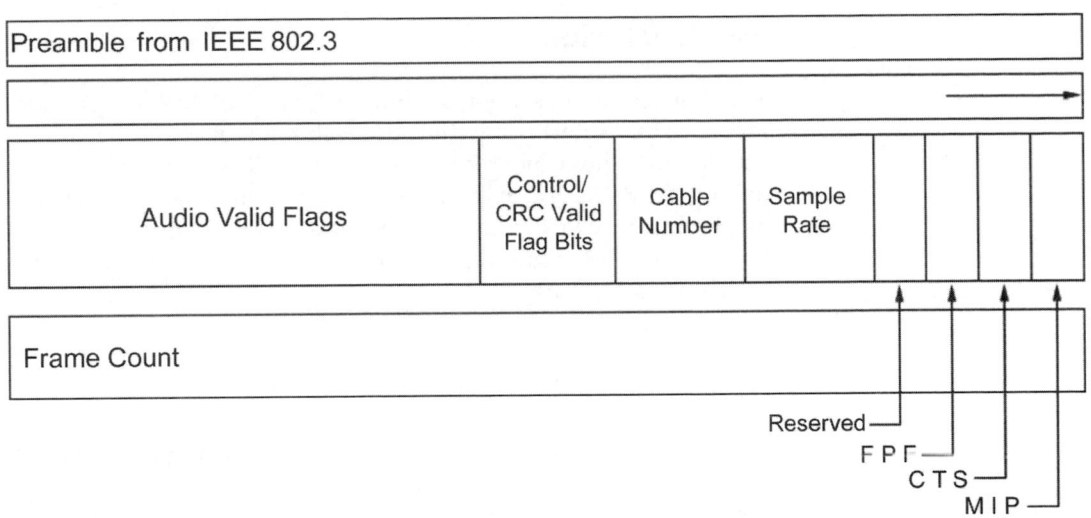

| Audio Valid Flags | Control/ CRC Valid Flag Bits | Cable Number | Sample Rate | | | | |

Frame Count

Reserved——
F P Γ——
C T S——
M I P——

Figure 5.6 GMICS frame header format. Source: *GMICS Engineering Specification*, Revision 1.1. (2000) Gibson Guitar Corporation.

the bus has cleared and transmission has begun (from the first packet). The new message cannot be sent until the passing of at least two frames in which the MIP bit has been reset.

Once a device has stopped sending a message, it must wait at least eight frames before it may participate again. In this way, each device has an equal chance of gaining control of the network for the purposes of transmitting its own message.

Addressing

Addresses are contained within a 48-bit address field, and to and from address fields are provided within the packet header. The address field is split into three fields of 16 bits. The number of devices on the network is therefore limited to 65 356.

5.5.5 Control pipe specification

The engineering specification also defines a method of carrying control information. The three 16-bit address fields contain information regarding the device address, the function address within the device, and the parameter address within the function. This method of addressing a control within a function in a device allows multiple functions to be located within a device, each with a number of uniquely addressable qualities (such as gain for instance).

The specification also allows other control protocols, such as MIDI, to be transported through the control pipe specification.

5.6 Conclusion

The interfaces presented within this chapter are proprietary, meaning that interfaces with other technologies rely on commercial forces. There are two exceptions to this, since GMICS is openly available for inspection and comment, and CobraNet is available for licence with the provision of a small printed circuit board, containing all the interface functionality required for inclusion in devices.

Notes and further reading

Alesis Corporation. 1633 26th Street, Santa Monica, CA 90404, USA. http://www.alesis.com.

BSS Audio. Linkside House, Summit Road, Potters Bar, Herts EN6 3JB, England. http://www.bss.co.uk.

Jusszkiewicz, Henry, Frantz, Richard, Flaks, Jason and Sherman, Thomas (2000) *Global Musical Instrument Communication Standard, Engineering Specification, Revision 1.1.* Copyright 1999. http://www.gmics.org.

MediaLink Product Overview (2000) Fairlight ESP Pty Ltd, Unit 2, 1 Skyline Place French's Forest, Sydney NSW 2086, Australia. http://www.fairlightesp.com.

Peak Audio, Inc. (2000) 1790 30th Street, Ste. 414, Boulder, CO 80301, USA. http://www.peakaudio.com.

United States Patent: 5 297 181 *Method and Apparatus for Providing a Digital Audio Interface Protocol.* http://www.uspto.gov/.

6 IEEE 1394 and Universal Serial Bus

Previous chapters have explained the basics of digital communication and illustrated concepts and implementations that have been proven over time. In Chapter 4, examples of communication techniques from the audio industry were then placed in the same context.

Both audio and computer communications technologies have advantages over the other, and one of the purposes of a book such as this is to compare and contrast the differences.

Broadly, audio communication interfaces have developed with a very stringent definition of the quality of data delivery that is required. Data communication interfaces, on the other hand, have developed with the ability to deliver data in a flexible and changing environment, with the result that the quality of service has suffered.

In this chapter, the first of several data transfer mechanisms is inspected in more detail. Each mechanism or interface differs from another, not only in the actual mechanism that is implemented for data transfer, but also in the limitations of the interface to which layers in the model are associated, and the quality of service on offer.

Hereafter interfaces are grouped together for discussion. The grouping is based upon technologies competing for implementation in the same areas which may be able to offer similar QoS, for instance. In this way, comparisons are more easily drawn.

This chapter presents IEEE 1394 (1394) and Universal Serial Bus (USB) for inspection. 1394 has been advanced for discussion in areas such as the transfer of data from hard drives over small distances to the CPU within a single small computer. On the other hand, 1394 has already developed into products that transfer data from computers to DVD players and televisions and back to computers, placing it as a connection interface for PC peripherals.

USB has also been developed as an interface to connect PC peripheral devices and so USB appears to occupy similar territory to 1394 in terms of the applications to which it can be put.

The possibility of commercial rivalry between the two technologies was avoided with Device Bay (1997), an initiative to improve the process of connecting devices such as hard disks, DVD, and CD players to the PC by using a combination of both 1394 and USB.

6.1 IEEE 1394

It is likely that the interface will become known by one or other trademarks such as i-link from Sony, FireWire from Apple, or mLAN from Yamaha (which is incidentally aimed at the professional audio market), depending upon which implementation is purchased by the consumer.

6.1.1 Administration

The central standard, around which the commercial work is based, is the internationally adopted standard known as ISO/IEC 13213, Information technology – Microprocessor systems – control and status registers (CSR) architecture for microcomputer buses (CSR Architecture). The document defines a common set of core features that can be implemented by a variety of buses.

Within the commercial sector, engineers tasked with investigating improvements to the Apple Desktop bus (ADB) focused on producing a fast, inexpensive, serial bus to replace ADB with an eye on replacing other ports by producing a new generic communications port on Apple's Macintosh workstation. Apple engineers adapted a flexible and inexpensive serial bus in the shape of the unimplemented IEEE 1394 standard, which describes a serial bus extension to the CSR Architecture.

Although originally intended for use in diagnosing errors on backplane parallel data bus, the standard was adapted in order to support the streaming of real-time data for multimedia purposes.

Commercial pressure from competing multimedia buses throughout the mid-1980s, combined with a difficult period for Apple research departments, resulted in the announcement from Apple in 1990 that the progress made in this direction would be turned over to public administration. Apple had already named the initiative FireWire at this point, and this remains as a familiar trademark. The public exposure resulted in innovations from other parties that increased the speed whilst keeping cost down, and operation simple.

By 1993, Apple had publicly demonstrated the ability to determine the quality of service by streaming time-sensitive data over the interface. In 1994, the 1394 Trade Association (see Notes and further reading) was formed to support and co-ordinate the development of computer and consumer electronics for the interface. The first compliant chipset was used in a digital camcorder implementation from Sony.

Unfortunately, the first ratified proposal, IEEE 1394–1995, contained enough inconsistencies to cause interoperability problems. This resulted in the IEEE 1394a supplement, which fixed these problems and added some extra functionality that the working committees had agreed in the meantime. One month after the supplement was approved in 1996, a group of companies approached the issues of audio streaming over 1394. The group included Microsoft, Texas Instruments and PAVO, a digital A/V product-development company (PAVO Inc. – see Notes and further reading). PAVO developed the audio implementation with Texas Instruments, working to integrate the 1394 chipset into the PAVO design. The result of the initiative was the 1394 Digital Audio Bridge, demonstrated by the Microsoft CEO, Bill Gates. The demonstration took place at the Windows Hardware Engineering Conference in 1996 (WinHEC'96), taking the form of a 1 minute Windows .WAV file sampled at 16-bit, 44.1 kHz resolution and transmitted from a PC hard drive to the auditorium sound system using a Texas Instruments PCI 1394 card. From the card, the signal passed through the PAVO Digital Audio Bridge before being transferred to an S/PDIF interface connected to the auditorium surround sound system.

6.1.2 Background

Early implementations of 1394 were able to provide data rates of up to 400 MB/s. This was far in excess of bandwidth offered by network techniques available in the same price bracket, such as the 10 MB/s and 100 MB/s options for Ethernet products, and so justified attention.

The principle of transferring packet-like computer network data, whilst maintaining deterministic QoS to real-time and interactive applications, was fundamental from the concepts inception.

Work continued on the standard, resulting in the 1394b supplement. The supplement ratifies data rates of 800 MB/s, 1.6 GB/s and 3.2 GB/s.

Several companies have embraced the development of 1394 since it was passed to the IEEE for administration and this has resulted in a number of different implementations or trademarks. These include Sony, Yamaha, Philips, Pioneer, IBM, and Adaptec. The work undertaken by these organizations effectively builds the next layer of functionality on top of those that have already been established and allows the interface to be applied in multitudinous circumstances.

6.1.3 The IEEE 1394 standards

Illustrated in Figure 6.1 is a conceptual model of the various interface implementations built upon the CSR Architecture. IEEE 1394–1995 uses the CSR Architecture as a foundation, in order to define serial bus implementations. Serial bus implementations are themselves further divided into two parts, backplane environment, and cable environment. The cable environments are in the context of inter-device data transfer.

As mentioned, the full IEEE 1394 standard is made up of a number of documents, the first of which is the core CSR architecture. On top of this is the IEEE 1394–1995 standard.

CSR architecture

The ISO/IEC 13213 standard defines the fundamental principles to be adopted by the bus architecture, including the node architecture, address space, common transaction types, interrupt and broadcast mechanisms and control and status registers.

The CSR architecture defines automatic configuration and various failure–recovery mechanisms, including power failure recovery by the operating system, and fault logging. The standards

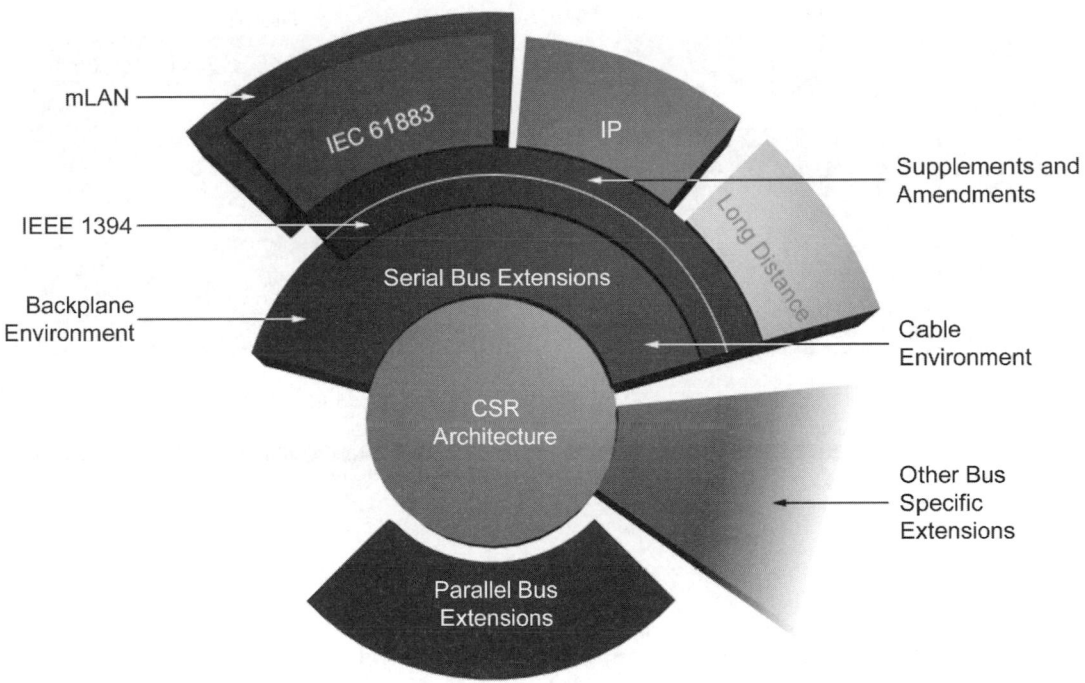

Figure 6.1 Conceptualization of IEEE 1394 standards framework, showing the CSR architecture as the core upon which other standards have been written. Once the IEEE 1394 standard was ratified, protocol definitions were then possible. Note the relationship between IEC 61883 and mLAN specifications.

framework permits bus-dependent extensions, and 1394 is such an extension.

A physical device attached to the bus is known as a module, and each module contains logical entities called nodes. A node in 1394 jargon is therefore different from a node in network jargon where a physical device is so termed. Instead, a node is an end point in a communication conversation and there may be multiple logical end points to a single cable attachment.

Each node is built up of functional subcomponents called units. Units identify a single resource such as processing blocks, memory services, and I/O functionality. Units may operate independently of each other and may share node registers or may be mapped into the node address space. Units are a little like the peripheral cards on a PC, and may be accessed by the operating system through a software driver. Figure 6.2 illustrates the relationship between modules, nodes, and units.

Nodes

Defined within the ISO/IEC 13213, nodes contain initialization code, CSRs, and ROM entries. The standard provides a framework to ease implementation on different platforms.

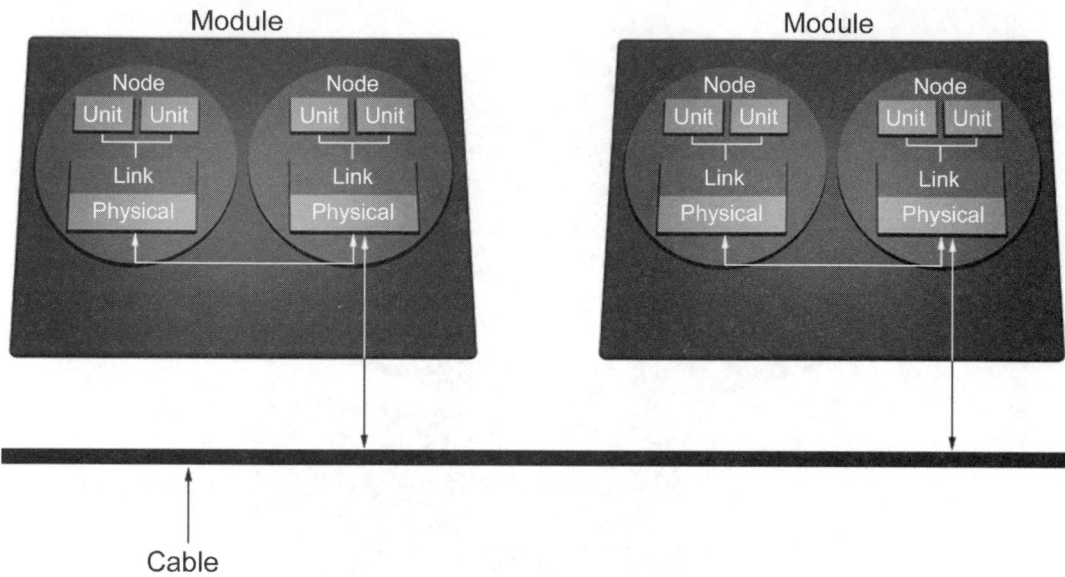

Cable

Figure 6.2 The relationship between modules, nodes and units in IEEE 1394. Within the standard documents, a physical device is called a module. A module may contain several nodes, each one performing a separate function. A node is made up of blocks called units, and each may be for processing, storage, memory, or capable of handling video or audio, for instance.

The ROM entries provide a 24-bit block containing the vendor identifier, a bit-information block, and the root directory. This root directory can be thought of as containing the boot code on a PC and is required when booting the interface.

6.1.4 General description

Probably the most significant aspect of 1394 in terms of the ability to transfer multiple audio channels to performance oriented service levels is the implementation of TDM in 1394. 1394 allows fixed rate multimedia streams to share the bus with packet-based computer data.

Time division multiplexing

The 1394 implementation of TDM specifies slices of 125 μs and a split in the bandwidth between asynchronous and iso-chronous transfers of 20–80%. As illustrated in Figure 6.3, all asynchronous communications on the bus are grouped together and guaranteed a minimum of 20% of the bandwidth and a maximum of 400 MB/s.

Physical layer definitions are ratified for several data rates, and the same thresholds apply to all the available cable/speed combinations specified in documents up to and including the 1394b supplement.

Figure 6.3 IEEE 1394 first TDM arrangement.

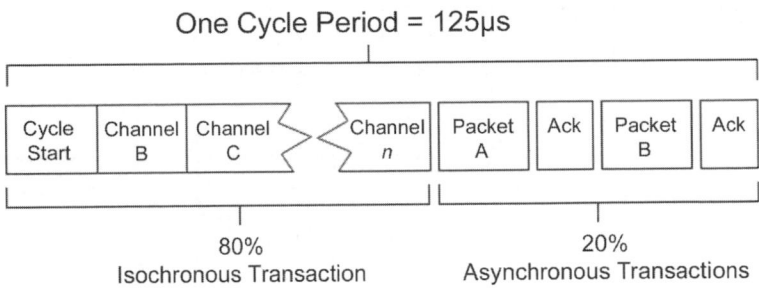

Fairness
========

Fairness

The asynchronous portion is further divided using a fairness algorithm. The fairness algorithm dictates that each asynchronous node wishing to broadcast onto the bus may perform exactly one transaction during each fairness interval. A fairness interval is described as complete when all nodes have been given the opportunity to perform one transaction.

In the isochronous portion of the bandwidth, broadcasts make an application before being allocated bandwidth. Applications are submitted to the device on the network known as the isochronous resource manager, which performs a similar role to the ringmaster of a token ring and is nominated as one node in one module on the bus. The function is one of three performed within the bus management layer of the standard and is covered in more detail in the following sections.

The communication model

IEEE 1394 defines its own communications model shown in Figure 6.4. This model still contains all the necessary functionality from the 7-layer model, but is organized slightly differently because of the two different types of transfer. The model illustrates that the communication of asynchronous and isochronous data types both require intervention from the bus management layer, but have different mechanisms for transmission and different obligations to fulfil. The methods reflect the 20/80 split in the bandwidth with the result that the concept requires the definition of two different but related communication models. Each draws from the other's functions wherever possible, in order to retain reciprocity, efficiency, and coherence between the types. The two mechanisms use as many common components of the bus management layer as possible, whilst remaining segregated within their asynchronous and isochronous time divisions on the bus.

Figure 6.4 IEEE 1394
communications model.

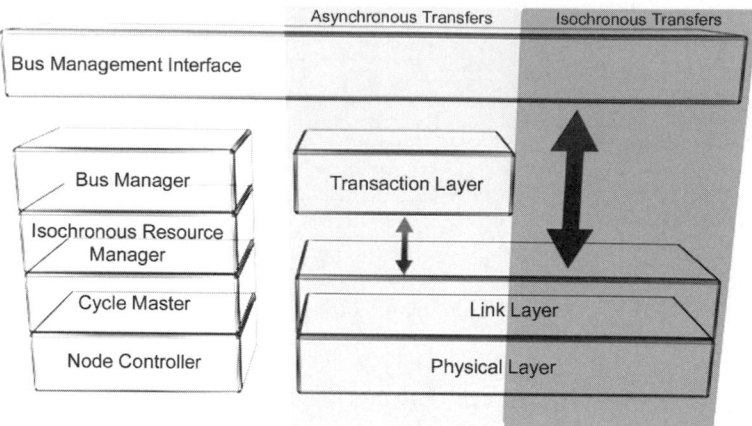

From the model, it is possible to see that isochronous transfers are slightly simpler to perform than asynchronous transfers, and the addressing scheme of the isochronous functions is simpler.

6.1.5 Physical layer – cabling, connectors and encoding

In order to simplify operation, IEEE 1394 devices attach to the bus without the need for resetting any other devices or the bus itself, in an elegant multi-access network. When a new device is attached to the network, the standard describes methods to continue normal operation, whether the new device is powered on or not.

Topology

Devices are cabled together in a similar fashion to physical MIDI cabling layouts, although the 1394 standard also allows support for hubs. Each 1394 compliant module contains one or more ports for attaching a cable. Cabling more than one device together begins a network, and such an example is shown in Figure 6.5.

Modules can be connected to as many other modules as there are ports and subsequent devices are attached via the spare ports. In this way the network is built up as a chain, in the same way as connecting MIDI in and out ports. Those modules with only one port cannot transfer information any further along the bus, since they can only be connected to one module.

The standard specifies 63 devices within a maximum of 16 cable hops and each cable is limited in length to 4.5 m, dependent upon

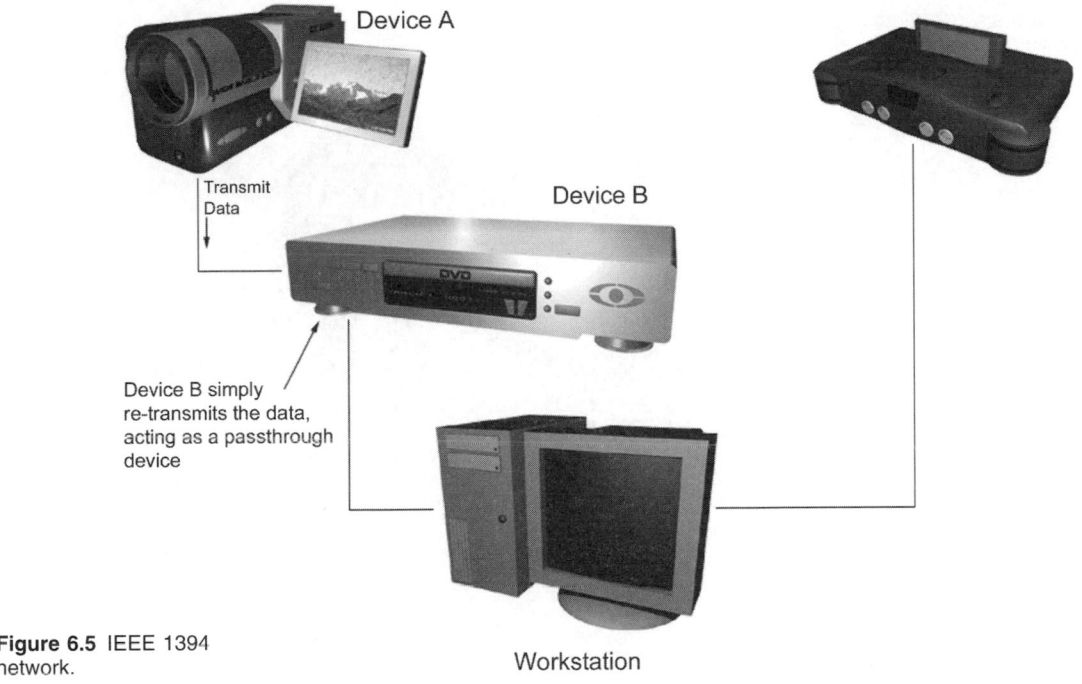

Device A

Transmit
Data

Device B

Device B simply
re-transmits the data,
acting as a passthrough
device

Figure 6.5 IEEE 1394
network.

Workstation

the cable characteristics, making it limited over distance.
Although the number of devices attached to the bus appears to be
quite small, the number of buses that can be connected together
is 1023. This might be thought of in the same way as subnetting
IP with a possible range of 1023 networks instead of 255 (one
octet) and a possible 63 devices on each subnet instead of the 255
available on class C IP networks.

Since the limitation of 16 hops would appear to make the
connection of over 64 000 devices difficult, the specification also
includes functionality for a bridge between buses. The bridge
allows connection in a physical star topology when using a multiport bridge, which is similar to a hub in function and appearance,
in that multiple ports are available for connection of devices.

Cable
The specification defines two cable pairs for the bus known as
Twisted Pair A and B (TPA and TPB). Both cable pairs are used to
transfer data and two-way communication is performed using
half-duplex operation, meaning that the same cable pairs will be
used for the return transmission.

Figure 6.6 IEEE 1394 cable construction.

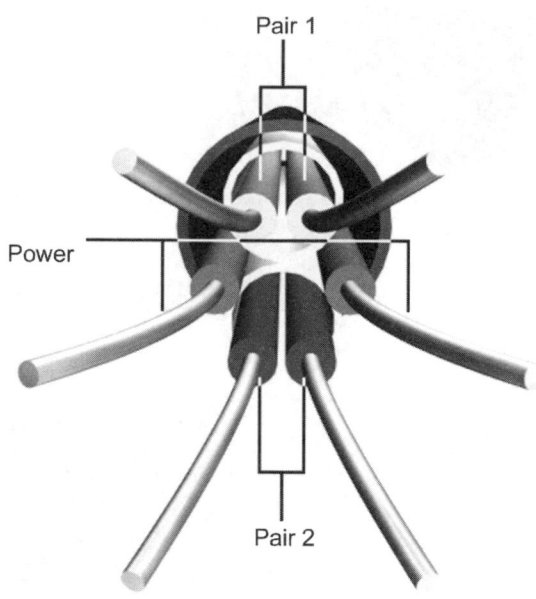

A section of IEEE 1394 cable is illustrated in Figure 6.6, and this is designed to be flexible and durable, with a diameter of less than 0.625 cm.

The physical connectors vary depending upon the implementation, for instance Sony uses a miniature version of the connector in laptop computers, known as i-link.

Power

1394 devices may be bus powered, receiving their power from an additional wire pair within the 1394 cable. This brings the total number of wires inside a 1394 cable to six, although the power pair may be dropped depending upon circumstances dictated by the 1394a supplement to the standard. The inclusion of power on the cable is intended for miniature devices and the power pair supplies electrical power in the range 8.0–33.0 volts (DC).

For 6-pin connections, the connectors also house the power pins. These are designed to make first contact when being plugged in, and last contact when pulled out, to ensure graceful entry and exit of the device onto the bus. The standard 1394 connector was originally created for the Nintendo Gameboy, and has proven to be reliable in the harshest environments (in the hands of children!).

Attachment

As per the 7-layer model, the physical layer provides the actual interface to the cable bus and the services required to do this are supplied by the PHY chip, which sits logically next to the cable, between it and the link layer services.

The first activity on the cable is the initialization process, during which the topology of the bus is discovered. This process involves each node talking to the others that are attached to it, until a root node, located at the logical centre of the bus, can be identified. Parent and child node relationships are built up and these are used during the arbitration process.

Once the bus is initialized, modules may begin transmitting data, but since 1394 is a multiple access mechanism, modules wishing to transmit data must first gain control of the bus through the arbitration process.

Arbitration

The arbitration mechanism for control of the bus begins after a period of idle time on the bus. The length of the idle period differs between isochronous and asynchronous transactions. For the isochronous portion, the idle time is between 0.04 µs and 0.05 µs, and for the asynchronous portion, the time can be tuned but must incorporate enough time to receive all parts of a transaction, as defined in the transaction layer. The arbitration signalling protocol is the same for both the isochronous and asynchronous portions of the interface, although the process itself is different.

Asynchronous arbitration

Within asynchronous transmissions, arbitration involves a node sending out a request packet to notify the root node that transmission is awaited. An acknowledgement packet is sent back to the device, indicating that the request has been received. The request–response pair takes place in a single transaction as these are considered to be molecular entities of the same transaction, meaning that arbitration of the bus does not need to occur for the response packet to be sent. This is an example of a multi-part transaction, for which the idle time must account.

If more than one node requests control of the bus at any one time, then the node closest to the root wins the contest. The controlling node can then transmit onto the bus following the period of idle time. Each node takes its turn to transmit data onto the bus, and does so once during the fairness interval.

Once ownership of the bus has been passed to a particular node, the PHY services within that node inform the link layer of the new status. The PHY receives data from the link layer, turning it into electrical properties on the cable itself. Since the data rate can vary, information about the speed of transmission is also sent within the packet transmission.

Isochronous arbitration

The isochronous approach to arbitration involves negotiation for a specific amount of bandwidth. The isochronous bandwidth is split into 125 μs periods, and bandwidth is allocated as a proportion of this.

Time periods are indicated on the network by cycle start packets transmitted by the root node. The root node is also defined as the cycle master, allowing it to automatically win any arbitration contest. The idle time for the isochronous portion is always smaller than for the asynchronous portion, resulting in asynchronous transmissions having a lower priority than isochronous transmissions. This prevents asynchronous data from interfering with the transmission of timely isochronous data.

From Figure 6.3, the cable is split into time divisions using the cycle start packets. Isochronous communication can take place directly into the divisions.

6.1.6 Link layer

The original 1394–1995 standard describes the interface between the PHY chip and the link layer components in the Appendix and is not a required implementation. However, the 1394a supplement specifies that the PHY–link interface is required if separate PHY and link components are used, in order to promote interoperability between components from different manufacturers.

The link layer sends a request to the PHY layer to send out a packet. The PHY responds by arbitrating for the bus and sending a grant signal to the link layer after control of the bus has been established. The link layer then interprets the grant signal as an indication that the PHY is ready to handle the packet, which is then forwarded. In some circumstances, the PHY layer may have reason to discard the request or the packet, and so the link layer also performs monitoring, so that it knows whether the PHY has discarded the packet. In this case, the link layer will begin the process again by reissuing the request upon the PHY services.

Physically, devices do not have to be attached in a particular order, as described for the signal path of digital audio data. The logical communications topology is point to point since any data requiring to go from device A to the workstation in Figure 6.5 passes through device B, which performs a similar function to a simple cable repeater. This technique is similar to that employed by Fibre Channel Arbitrated Loop topologies (see Chapter 7).

Both asynchronous and isochronous packets come through the link layer, although each has a different addressing scheme and packet structure, and so are easily identifiable from each other.

Encoding

For data transmission, data are encoded onto the bus using a combination of NRZ and data strobe encoding. Data are transmitted on TPB with the strobe on TPA. This method of encoding allows the clock signal to be removed by performing a logical exclusive OR (XOR) on the data and strobe signals.

6.1.7 Transaction layer

The transaction layer is important to the interface since 1394 is essentially a transaction-based communications interface. Trans-actions are split into their familiar 1394 packet types of asynchronous and isochronous, and each type has its own addressing scheme.

Addressing

Isochronous transactions use a channel number addressing scheme similar to that employed by MIDI and these channel numbers are assigned automatically when the bus is reset. This is possible since isochronous packets are essentially broadcast onto the network and picked up by one or more receivers. Isochronous transfers do not require any response from the receiver, since any action that might normally be taken as a response to lost or damaged packets will occur too late for any remedial action to occur.

Error detection

The implementation of error checking places a CRC field at the end of the packet (see Figure 6.7). Any error checking mechanism reliant on resending data is considered redundant, since the

Asynchronous Packet Structure

Destination Address	Source ID	Transaction Type	Transaction Dependent Data	CRC

Isochronous Packet Structure

Channel Number	Transaction Type	Data	CRC

Figure 6.7 IEEE 1394 packet structure.

isochronous stream would need to be buffered to allow the corrections to be made before playing out correctly over time and so synchronization with other streams would be compromised. This type of error checking is therefore removed from the isochronous portion.

The transaction layer can now be by-passed, as the purpose of the transaction layer is to control end to end conversations and ensure accuracy of asynchronous data transfers through the use of acknowledgement packets. As a result, the matter of setting up and closing down the isochronous transaction is left to the link and PHY layers.

A system of acknowledgement packets ensures that transactions are successful. If the acknowledgement packet is not received in a timely fashion, or if the packet indicates that a CRC error was detected in the original transmission, then the response is application dependent, usually resulting in retrying the transmission.

Asynchronous transactions

Asynchronous communication uses hardware level addressing that can be considered identical to MAC addresses familiar from other types of computer network.

Asynchronous transactions essentially occur in four stages, as shown in Figure 6.8. The first stage is that a node requiring an action from another node will send out a request packet. The receiving node then returns an acknowledgement packet in response. Although the acknowledgement packet is a separate transmission, they are defined as molecular parts of the same request and so as before, arbitration need not occur.

The third part of the transaction is performed by the receiver, which sends out the requested data. Upon successful transmission, another acknowledgement packet is sent, confirming the safe arrival of the data. The acknowledgement packet follows the same rules of arbitration as before.

If the request for data can be fulfilled quickly, then the device may be able to send the acknowledgement packet along with the data that was requested. This is known as a unified transaction and is designed to improve the efficiency of the bus. However, if the request cannot be fulfilled quickly enough, then the acknowledgement packet is sent out before the data request has been fulfilled. In this case, two complete transactions occur.

The first transaction is the original request plus the acknowledgement, and the second transaction contains the data travelling in the other direction, plus the acknowledgement. This is known as a split transaction, and each part of the transaction must arbitrate for control of the bus before sending data.

Figure 6.8 Four stages of asynchronous communication in IEEE 1394. Stage 1, a requesting device (A) sends out a request packet to B. Stage 2: device B responds by acknowledging receipt of the request. Stage 3: device B sends the requested data. Stage 4: the original requestor acknowledges receipt of the data.

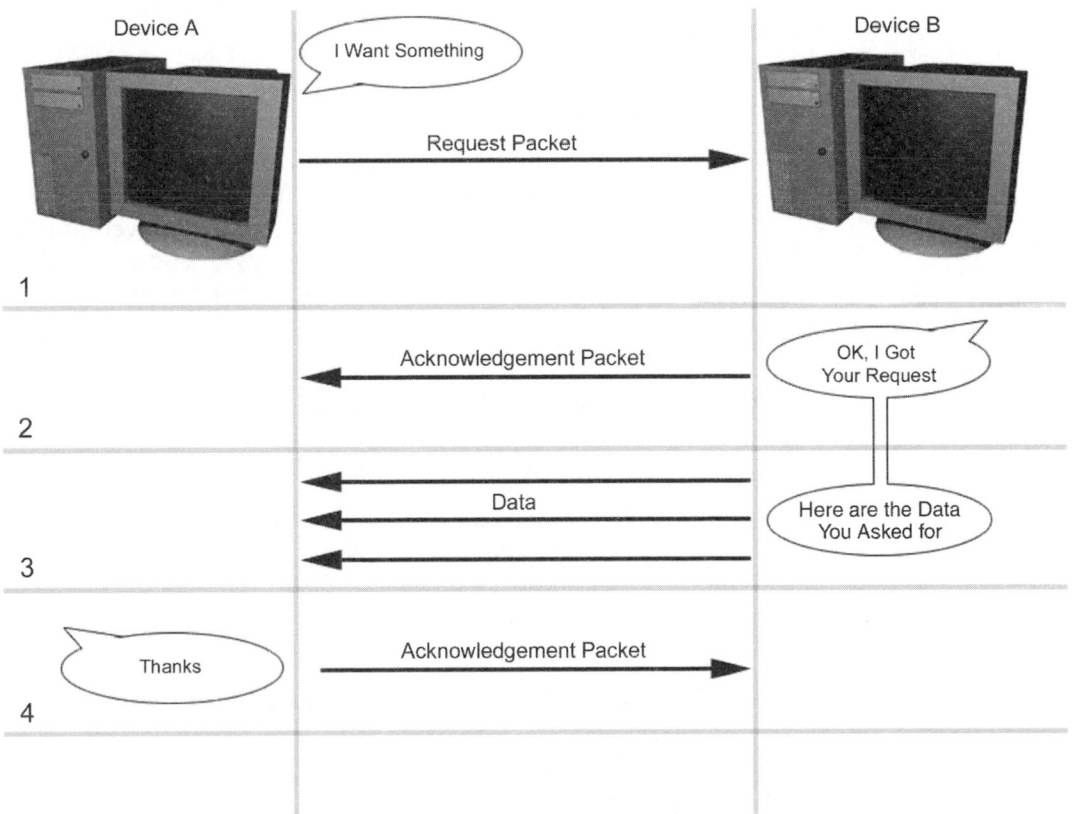

Isochronous transactions

The isochronous portion uses a channel number to identify target nodes instead of the node ID and device address, as in the asynchronous portion. The isochronous resource manager assigns the channel number during the initialization. The number is generated from a 64-bit register which acts as a map in which each bit represents a channel number.

The isochronous transaction is much simplified compared to the asynchronous portion, and contains only a request data phase, with no acknowledgements and no response.

Bandwidth is requested by the sender and allocated from within the IRM register known as BANDWIDTH_AVAILABLE. This contains information on the total bandwidth available on the network. The register splits the bandwidth into reserved and remaining units, and requests are fulfilled from the remaining units.

Isochronous transactions use the data block packet to frame data and this frame type was originally not applicable to the asynchronous time division of the interface, although support was added later.

6.1.8 1394 extensions

As mentioned at the beginning of this chapter, 1394 has been developed in the context of many different technologies and industries. These developments take the form of layers further up the communications model, and deal with subjects such as the format of data within one or other portion of the interface.

For instance, it is necessary for nodes concerned with the transmission of audio data over the interface to have a common understanding of the format of the audio data in order that correctly interpretation is achieved.

1394 does not make any specifications as to the format of data, and so upper layer protocols need to be determined. As such, many working parties have been set up to address the various needs of groups expressing an interest in the use of IEEE 1394.

6.1.9 Audio and music data transmission protocol

A contender for the standard most relevant to digital audio data and 1394 is IEC 61883, specifying a protocol designed for the transmission of streamed audio over the interface.

The standard consists of six parts, intended as audio/video specific extensions to 1394 and is variously entitled 'IEC 61883 consumer audio/video equipment – digital interface'.

'Part 1 – General' is the subtitle given to the first document in the series. Part 1 describes the general packet format, data-flow management and connection management for audio/visual data, and also the general transmission rules for control commands. The objective of the document is to define the transmission protocol for audio/visual data and control commands that provide connectivity between digital audio and video equipment using the 1394 standard. The positioning of IEC 61883 (61883) in relation to the 1394 documents is reinforced in section 4, which states that all the physical, link, transaction and management layer functionality remain untouched by the 61883 extensions. Any implementation of the IEC extension must include IRM capabilities within the module and most importantly also includes an implementation called plug control registers, detailed in section 7 of IEC 61883–1.

Plug control registers (PCRs) are special purpose CSRs and under the PCR model any device capable of transmitting or receiving isochronous data implements virtual plugs for its inputs and outputs. A connection is made between an output plug on one node and one or more input plugs on other nodes. PCRs on each node specify the isochronous channel, device specific I/O port numbers, and other parameters necessary for managing virtual connections across the bus. Asynchronous transactions called connection management procedures (CMP) are used to communicate PCR settings. In this way, more complex signal routing can be designed within a single or multiple modules.

The PCR mechanism essentially allows audio processing blocks to be connected together in the same way that audio effects processors are patched in the analogue domain. In the case of 61883, the connection is logical and made within the unit itself, but the concept can be extended to include other units on the same bus, and so powerful audio processing paths can be built. PCR settings may or may not be persistent when a device's power is cycled, depending on the implementation.

IEC 61883 also defines another set of connectors specifically for A/V consumer equipment, to further add flexibility to the 1394 standard suite.

Yamaha originally proposed the definition of the audio and music data transmission protocol (A/M protocol) described above, as a result of work on their mLAN initiative. The mLAN implementation of 1394 is aimed specifically at professional audio applications. The A/M protocol was ratified by the IEC and can be applied to all modules or devices that have any kind of audio and/or music data processing, generation and conversion function blocks.

The protocol is designed to accommodate all the information contained within IEC 958 (AES3 and S/PDIF) packets and so enhances compatibility with other professional and consumer interfaces. The specification also allows the transport of raw audio samples and MIDI data, and its conceptual position is shown in relation to other parts of the standard in Figure 6.1.

6.1.10 IP over 1394

The extension to 1394 that is perhaps most illustrative in terms of the 7-layer model is the work designed to allow the use of the IP protocol over 1394. Although the interface allows asynchronous packet switched data to be transmitted, it makes no definition as to how this is to be implemented.

The IP1394 working group avoids reinventing the wheel by selecting a well-known and understood solution to the upper layer processing in the shape of the Internetworking Protocol.

The work of IP1394 specifies how to use the IEEE 1394–1995 interface for the transport of Internet Protocol version 4 (IPv4). This initial proposal defines the methods, data structures and codes, as well as the more difficult business of proposing an address resolution protocol between IP and the node id and device address for specific use on the serial bus. On the Internet, this is performed by the address resolution protocol (ARP) which identifies the MAC address that a particular IP address is using to allow the physical layer transmission to occur. This stage takes place at the last router hop before reaching the destination device.

The proposal identifies that not all modules on a serial bus can accept IP datagrams and makes allowances for this. Those devices that are IP enabled must have the ability to be a bus manager and be capable of sending and receiving asynchronous packets.

IP packets are encapsulated within the 1394 frame structure, as part of the core load, and will be split up into fragments where the IP packet is too big for the maximum packet size dictated by 1394. Essentially then, IP is enclosed within 1394 handles, and sits on top of the 1394 functionality within any reference model (IP1394 Working Group – see Notes and further reading).

There are no short-term plans to introduce IPv6 over 1394, allowing support for the increased address space and additional functionality of the later version of IP.

6.1.11 1394 long distance

Although 1394 would appear to be a perfect technology to solve the audio connection conundrum, one of the limitations of 1394 is the distance over which it can run. Cables are specified at a maximum length of 4.5 m, making it a suitable cluster networking technology for groups of computers working together using a fast link, but little use in multiple access networks, where longer cable runs are normally anticipated.

The distance limitation is an obvious candidate for improvement and two early successes have been noted. The first of these was a real-time digital video link between the United Sates and Japan (Keio University – see Notes and further reading), demonstrated during the Super Computers 1998 exhibition in Orlando, Florida, in November of that year. The link utilized 1394 connected to both ends of an Internet service link supplied by TransPAC.

During the event, the 35 MB/s Internet service proved inadequate, even though the link was made private from the rest of the Internet. At the time, it was calculated that 40 MB/s would be the appropriate bandwidth to perform the operation with complete success, and an algorithm to discard frames was also considered for lower bandwidth applications.

The second notable event in the context of the distance limitations of 1394 was taken by the Leviton Telecom company (see Notes and further reading) in the United States, who worked on an NEC transceiver to produce their range extending module (RXM). This was demonstrated at the winter Consumer Electronics Show in Las Vegas during 1998. The system was still in the test phase of development, pending the outcome of the 1394b supplement by the IEEE 1394 working group. The intention of the initiative was to extend the network throughout a single building for residential applications.

6.1.12 Other extensions

Further extensions to the 1394 interface include its application within LANs (FireNet), as well as Digital TV, and other consumer electronics. In addition, an important progression is the adoption by the AES of the subject of professional audio over 1394, in the AES-X95 initiative as covered by the SC-06-02 committee.

The scope of this project includes the inspection of 1394, 1394a and IEC 61883-6 PAS, with the intention of identifying additions required for the transfer of professional audio over 1394, including proposals for jitter, channel assignment, MIDI transfer and synchronization (AES, SC-06-02 – see Notes and further reading).

6.1.13 Conclusion to IEEE 1394

The IEEE 1394 interface offers two logical network types over a single physical cable. The networks are separated into time slices and do not interfere with one another, except where this may be designed into the application within the upper layers.

The standard is a result of work performed on many documents and development appears to be slow, principally because of the sheer numbers of possible applications that the bus could be put to. The work has resulted in early development products, such as compliant semiconductors and integrated circuit boards, as well as commercially available consumer products such as camcorders and backup devices from various manufacturers.

IEEE 1394 consumer goods appeared on the shelves as early as 1998, although caution and thoroughness meant that the promised interoperability was not necessarily part of the interface. For instance, early implementations of i-link within Panasonic camcorders only offered digital video output from the i-link port, with no capability for input (Panasonic NV-DS33EG digital video camera DS33, and others).

However, the core standards have withstood inspection by a disparate range of industries and some confidence can be assigned to the possibilities. The technology will become popular in the home, simply because few manufacturers are likely to ignore the competitive edge that network capability offers on consumer electronics. In the meantime, comprehensive work is being performed to complete professional standards capable of using the interface in such areas as audio and video.

6.2 Universal Serial Bus

USB was originally developed in 1995 in an initiative involving Compaq, Microsoft, Intel, and NEC. The motivation behind the initiative came from three interrelated factors, the first being the improved connection between telephone and PC. The other factors were the recognition of the inflexibility in configuration of the PC and the requirement to have a new interface for each peripheral. In other words, the motivation came from the lack of a Universal Serial Bus (Universal Serial Bus Specification, 1998a).

The specification describes the bus attributes, the protocol definition, types of transactions, bus management, and the programming interface required to design and build systems and peripherals that interoperate in an open architecture (Universal Serial Bus Specification, 1998b).

The specification was intended as an enhancement to the PC architecture, spanning portable, business desktop, and home environments with particular focus on computer telephone integration (CTI).

The initiative intended to produce a flexible specification to avoid problems of backward and forward compatibility. The resulting recommendations described a low-cost, ease of use platform capable of quick deployment for general-purpose communication between a PC and a peripheral device.

The USB 2.0 specification increased the speed of the interface to 400 Mbits/s, thereby turning in performance within range of 1394, and allowing the connection of a wider variety of peripherals. The interface was quickly supplied as standard in most new PCs, meaning a greater acceptance in the marketplace than 1394 initially achieved.

6.2.1 General description

The interface is based upon a master–slave arrangement, whereby the PC host operates as the master, with the peripherals obeying a defined protocol. The behaviour of the interface is described through several layers of the 7-layer model, and every communication over the cable is first negotiated through the host.

Similar to 1394, USB also manages the dynamic (on-line and real-time) attachment and detachment of devices to the bus. This phase involves identifying the new peripheral, and loading the required driver within the PC master, if one is not already loaded. Once the identity of the new device has been discovered, a unique address is assigned to the device during the enumeration process.

6.2.2 USB operation

Once the master and slave are configured for operation, the master initiates transactions on the bus to specific peripherals, which respond accordingly.

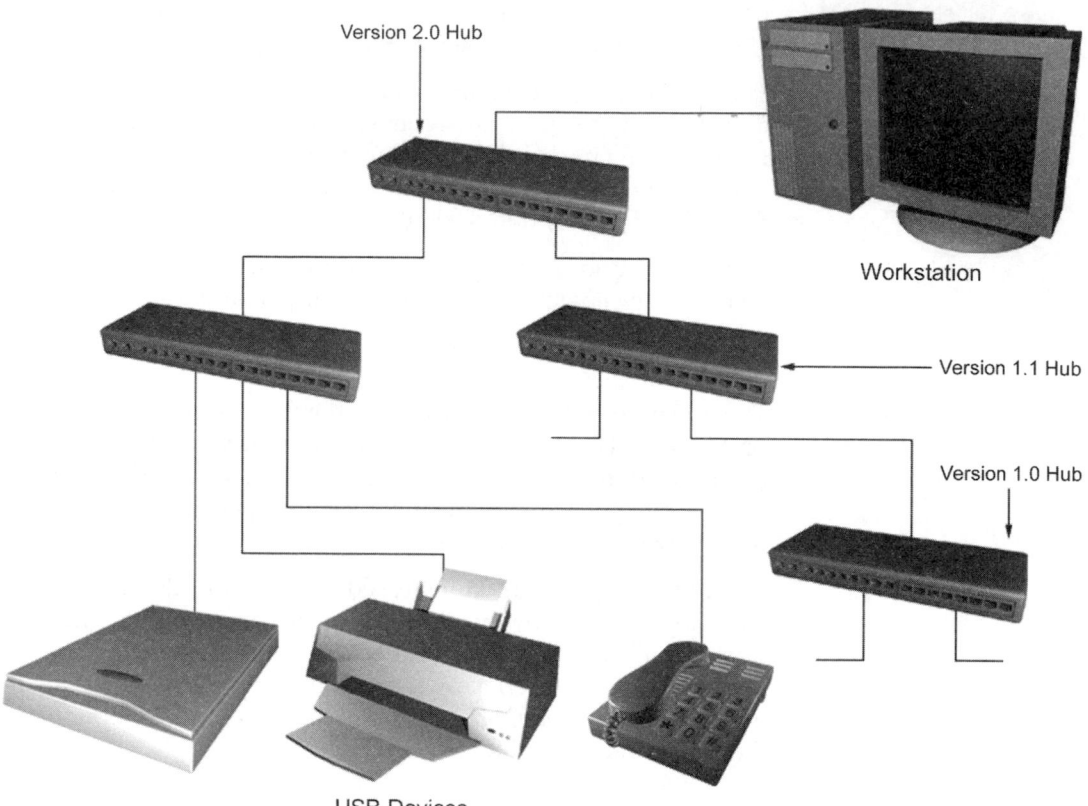

Version 2.0 Hub

Workstation

Version 1.1 Hub

Version 1.0 Hub

USB Devices

Figure 6.9 USB allows backward compatibility. The host is expected to remain up to date, but equipment adhering to older specifications is attached to newer versions and so communication always occurs through the newer hub. In this way, devices adhering to newer and faster versions of the specification are not hindered in providing the best possible communications speed by connecting through older equipment.

In order to make USB 2.0 backward compatible with previous versions of the specification, USB 2.0 specifies a micro-frame of one-eighth of a 1 ms frame. The updated and improved USB 2.0 specification implements the promised backward compatibility, and both versions of the specification can co-exist on the same cable, with individual data rate negotiations occurring for each device on the bus. Devices complying with older versions of the standard can be separated by means of a hub, and placed further away from the master in the tree cabling structure, so as not to incur any performance penalties for faster devices located closer to the host, as shown in Figure 6.9.

Specific mention is made within the standard of isochronous transfer of data with guaranteed latency, particularly audio, in the context of telephone integration.

6.2.3 Components

A USB interface consists of three broadly defined components: the peripheral device or slave, the PC master or host, and the USB

Figure 6.10 USB topological concept model, showing the tiered effect of device connection.

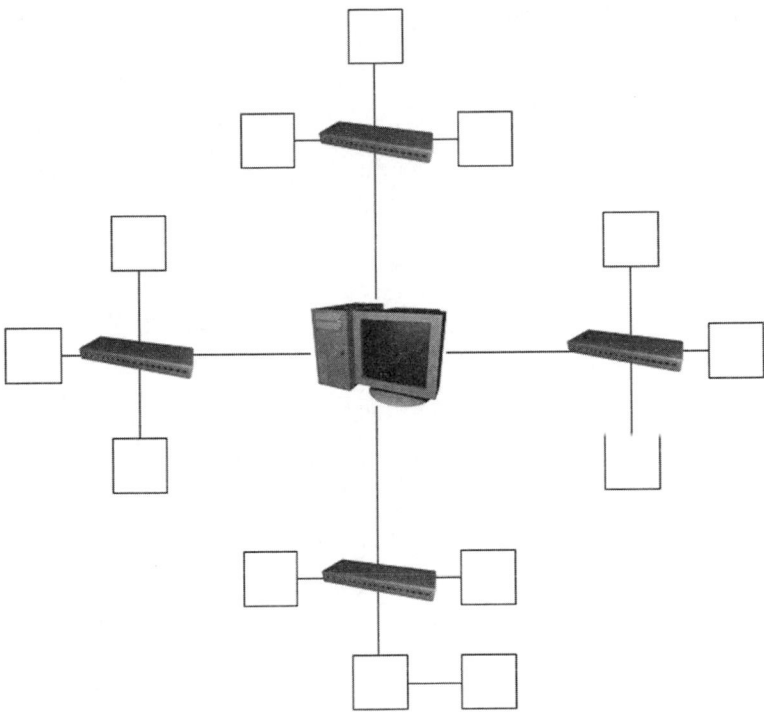

interconnect. The interface is further defined as consisting of the bus topology, the interlayer relationships (from the USB capability stack model, rather than the 7-layer model), data flow models and the USB schedule.

The bus topology consists of a connection model between USB devices and the host, as shown in Figure 6.10. The interlayer relationships are USB tasks performed at each layer in the system. Data flow models describe the manner in which data are transferred between devices and the host, and the USB schedule supports isochronous data transmission by eliminating arbitration overhead through scheduling techniques.

6.2.4 Physical layer

Topology
The cable topology is described as a tiered star topology because of its schematic representation shown in Figure 6.11. This is very similar to a normal star topology, except that a hub can be cascaded from another hub, thereby creating a new tier. This technique is sometimes employed in Ethernet networks, although Ethernet hubs often specify a particular port on the hub

Figure 6.11 Representation of tiered star topology as implemented within USB. Source: *Universal Serial Bus Specification, Revision 1.1 USB Cable Construction.*

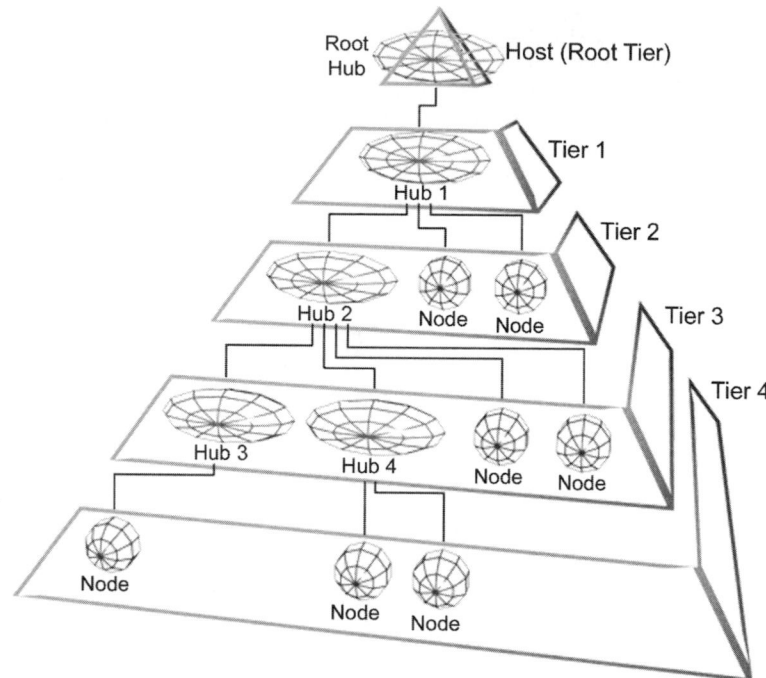

that allows cascading, whereas in USB, cascaded connections are no different from device connections.

The ability to cable in this way is supported by the USB arbitration behaviour, so becoming invisible to the device.

USB defines that devices must connect through a hub, since the hub performs a role in the arbitration mechanism, although in practice devices can be chained directly from the host.

Cable and interconnects

The USB cable contains four wires, two of which are in the fashion of a twisted pair and are used to send data, as shown in Figure 6.12. The remaining two are used to deliver power over the bus. Cables have different conduction definitions to allow for loss in signal strength, depending upon the length.

Two connection types are defined: an upstream connection required by all devices (shown in Figure 6.13) and a downstream connection. These are physically different from one another in order to prevent accidental illegal connection.

Encoding

Electrically, data are represented in binary form by varying voltages, depending upon the load, and the data rate can be

Figure 6.12 USB cable.

Figure 6.13 USB connector.

switched dynamically between 1.5 MB/s and 12 MB/s (USB 1.1 specification). Data are encoded with the clock signal using NRZI.

Power

Each USB segment provides power over the cable. The host supplies power for use by USB devices that are directly connected. USB devices powered from the cable are called bus-powered devices and those that have an alternate source of power are called self-powered devices. A hub can also supply power to USB devices connected to it, although some allowances are made for bus-powered hubs.

The standard also specifies the ability to use a power management system which is independent of the USB. The USB system software interacts with the host power management system to handle system power events such as suspend or resume.

Error detection

USB makes the assumption that the error rate of the transmission medium is of a similar nature to that of a backplane, in that any errors are transient in nature. Error protection takes the form of a data field within the frame, which can be implemented depending upon the service requirements of the data type.

The physical layer protocol includes separate CRCs for control and data fields of each packet. A failed CRC indicates a corrupted packet and the CRC can repair all single- and double-bit errors.

Attachment

Like 1394, USB also supports the dynamic attachment and removal of devices to the hub. The action of removal or attachment alters a status indicator within the hub. The host periodically queries the hub to retrieve the status indicators, and when an attachment is identified, the host instructs the hub to enable the port for data transfer. The host establishes control of the device through a default address and the device is then assigned a unique address, through which all subsequent communication occurs.

If a device is removed from the hub, the hub closes the port and changes the status indicator. Upon collection of the changed status indicator, the upper layer interfaces within the host are closed in an orderly manner.

Transmission

Whereas a physical AES24 connection may present more than one logical node to the cable, IEEE 1394 presents one or more nodes containing one or more logical units to the cable. Therefore, USB devices may present multiple endpoints which are uniquely addressed within the device. Data are transferred between the host and an endpoint and such conversations are known as a pipe.

Transmissions are classified into four types in the USB specification. These are control transfers, bulk data transfers, interrupt data transfers and isochronous data transfers.

Control transfers are used to configure and control a device. Bulk data transfers are transmitted in relatively large and bursty quantities such as might be used with printers and scanners. Interrupt data transfers are used for basic interaction devices, such as screen, keyboard, and mouse. Isochronous data transfers are also called 'streaming real-time transfers' and encompass QoS requirements for digital audio data.

In the case of isochronous transmissions, timing is implied by the rate of transmission, which in the case of PCM audio data is dependent upon resolution.

Buffering

Latency depends upon the buffering capabilities of each endpoint and error checking is removed from the physical layer in order to support the QoS requirements.

Clock

At the physical layer, the specification describes a bus clock that runs at a slower time than the sample clock. This clock depends upon the rate of transmission of start of frame packets and runs slower than the sampling rate, at 1 kHz. The natural time is otherwise determined by the data rate of samples moving on the interface.

6.2.5 Data link layer

Transactions

The USB network is a polled bus wherein the host controller initiates all data transfers by periodically sending out a packet describing the type and direction of the transaction. This packet is referred to as the token packet but its functionality should not be confused with that of the token ring token. All transactions involve the transmission of up to three packets.

Pipes

The endpoint acts as the data receptacle and endpoints converse with the host through a communications pipe or channel. All devices support at least one pipe through which the host may communicate and this is known as the default control pipe, used at initialization.

The earliest point at which a pipe can be allocated bandwidth on the interface is when the pipe is established. USB devices are required to provide some buffering of data. It is assumed that USB devices requiring more bandwidth are capable of providing larger buffers. The satisfaction of QoS requirements for iso-chronous transmissions is bounded by the latency and depends on the packet size.

Additionally, pipes have associations of data bandwidth, transfer service type, and endpoint attributes. The maximum packet size and bandwidth constraints are supplied as endpoint attributes during the discovery process. The same is also true for buffer size, which dictates the latency, since the size of the buffer determines the amount of data that can be assembled before being accessed or forwarded.

Data are addressed to a particular endpoint on the interface, using the device address (ADDR) and endpoint (ENDP) fields, which are found in the frame token. The recipient USB device identifies a relevant transmission by decoding the appropriate address fields with the frame.

In a given transaction, data are transferred either from the host to a device or from a device to the host. The direction of data transfer is specified in the token packet. The source of the transaction then sends a data packet or indicates it has no data to transfer. The destination then returns an acknowledgement packet, which indicates whether the transfer was successful.

Data format

Data transported by message pipes is carried in USB-defined structure, but the USB allows device-specific data structures to be transported within the USB-defined message data payload, by making no definition of the format of the payload.

The standard specifies two types of pipe, and these are stream and message. Stream data have no USB-defined structure although the transaction schedule allows flow control for some pipes. The flow control mechanism can manage flexible sched-ules, accommodating concurrent servicing of a mixture of stream pipe types, thereby allowing multiple stream pipes to be serviced at different intervals and with different packet sizes.

Isochronous transfers

Again like 1394, USB uses a transaction-based access method and the transaction mechanisms are defined for each of the four supportable transfer types. Each type of transaction has elements in common with other types, such as being initiated from the host by sending a request to a device to send data. If data are available, they are immediately sent within a data frame. If the data frame is empty, then no data were available to send.

Data structure

Although no data structure is described within the USB specification for isochronous transfers, one section is entitled *Special Considerations for Isochronous Transfers* (Universal Serial Bus Specification, 1998c). Within this section, a framework provides for the implementation of isochronous devices, defining synchronization types, the method for isochronous endpoints to provide data rate feedback, and how they can be connected together.

Synchronization type classifies an endpoint according to its ability to synchronize its data rate to the data rate of the endpoint to which it is connected. Feedback is provided by indicating accurately what the required data rate is, relative to the frequency of the start of frame transmissions. The ability to make a connection between endpoints depends on the quality of connection that is required, the endpoint synchronization type, and the capabilities of the host application that is making the connection.

6.2.6 Audio over USB

Devices classes

In the same way that 1394 builds around the core CSR architecture standard, so USB uses the 1.0 specification on which to build additional functionality. The USB interface provides a mechanism almost entirely within the physical and data link layers, with descriptions of endpoints to which data conversations can be attached.

Although USB 2.0 extends the functionality of the original specification, related standards and descriptions of device classes document agreed functionality further up the layers.

As mentioned, the USB architecture makes no assumption regarding data structures, although a means of extensibility is included through the definition of device classes. Further descriptions of the format of the data are left to these definitions of device classes.

Device classes are specified in separate documents known as device class specifications. Documents exist describing an audio device class (Universal Serial Bus, 1998a), and audio data formats (Universal Serial Bus, 1998b).

Functions

A function is a USB device that is able to transmit or receive data or control information over the bus. A function is described as being an implementation of a peripheral device with a cable plugged into a port on a hub.

However, a physical device may implement multiple functions and an embedded hub with a single USB cable. Such devices are known as compound devices and these appear to the host as a hub with one or more non-removable USB devices. A function then is a logical device where more than one function can exist in a single physical device.

Each function contains configuration information describing its capabilities and resource requirements. The host must configure a function before it can be used and the process of configuration includes the allocation of bandwidth and selection of function-specific configuration options.

The type of message stream defined by the function is dependent upon the class of device.

Audio devices

The audio device class definition (Universal Serial Bus, 1998c) applies to all devices or functions embedded in composite devices that are used to manipulate audio or contain audio-related functions. However, the handling of machine control such as for tape machines is specifically not covered, although MIDI data are considered as being directly related to audio data, and therefore are included within the specification. Furthermore, a class definition exists for MIDI devices (Universal Serial Bus, 1999).

The audio device class adheres to the synchronization mechanisms described within the USB specification and includes no additional information regarding the transport of data.

Three subclasses of the audio interface class are defined for the purpose of controlling audio functions. These are audiocontrol interface (AC), audiostreaming interface (AS), and a MIDIstreaming interface (MS). A combination of these logical interfaces will contain one AC, and any number of AS, and MS logical interfaces grouped together into an audio interface collection (AIC).

Audio functions are broken down into units and terminals. Units provide the building blocks to describe audio functions which, conversely, are built using a number of audio units thus providing a complete description of the device functionality. A terminal is associated with an audio data transfer, and is generally used to generate or sink audio data.

Seven types of units are described:

- Input terminal
- Output terminal
- Mixer unit

- Selector unit
- Feature unit
- Processing unit
- Extension unit

These units further extend the hierarchical concept of the specification by containing descriptions of functionality called audio controls. Each control has a set of attributes made up from the following:

- Current setting
- Minimum setting
- Maximum setting
- Resolution
- Memory space

The audio device class definition uses the example of a volume control within a feature unit as an example. The host requesting the attributes for this control may receive the current setting, allowing the correct display on a computer screen, along with the minimum and maximum settings, allowing the limitations to be understood, and a scale to be correctly drawn, for instance.

Taking the example of a mixer unit, a number of logical input channels are transformed into a number of logical output channels. Input channels are grouped together into clusters, each cluster entering the mixer unit through a physical input pin. After the mix processing, output channels are grouped together, with each cluster leaving through an output pin.

By creating processing blocks in this way, complex functionality can be built up from simple components in the same way as 1394 uses the configuration of nodes and units to achieve a similar result.

A feature unit provides processing on audio channels, and a processing unit represents a functional block inside an audio function, providing one of several defined transformations (including proprietary transformations) of the audio data, such as AC-3 (Dolby 5.1), 3D-stereo extending, reverberation, chorus, noise removal and dynamic processing, for example.

Finally, an extension unit provides vendor-specific extensions to the specification.

Data formats

Within the audio device class specification, data formats are described as being determined within a code located in the wFormatTag field of the class-specific interface descriptor.

Detailed descriptions of the data format are left to recommendations within further standards.

The main document for reference in this area is the Universal Serial Bus Device Class Definition for Audio Data Formats document (as referenced). The data format document classifies three types of audio data, named simply Type I, Type II and Type III.

Type I defines audio data streams constructed on a sample by sample basis, so that PCM words are sent constantly over time. Type II defines audio data types that do not retain any channel information during transmission and Type III defines audio data formats using a combination of Type I and Type II techniques, with the transmission rules similar to Type I.

The basic structure used to transmit digital audio data is the audio subframe, which contains a single audio sample. The size of the subframe is indicated in the bSubframeSize field and can be up to 4 bytes. The resolution of the sample, given in the bBitResolution field, must remain the same throughout the transmission, as must the channel order of the sample within the USB packet.

6.3 Conclusion

Both 1394 and USB have been developed to solve different areas of interconnection. USB is specifically aimed at connection to the PC, whereas 1394 is more generally considered for wider application, especially within residential spaces. The direction of the commercial market does not preclude either USB or 1394 from wider application.

The extended versions of each specification, USB 2.0 and 1394b follow the normal model for progress, in that more and better features are offered for the same cost. In this case, the only comparable product is that of a network access product, which compares poorly when considering quality of service and data rate. Comparable networking technologies include ARCNet, Ethernet, and token ring.

Both USB and 1394 are capable of interconnecting devices for general-purpose communication and with the data rates offered, these interfaces are attractive answers for communicating data. However, both also have limitations preventing them from being adopted as a computer networking solution, such as the maximum cable length and so on.

It is possible to speculate that the technologies could be improved upon through research, in order to overcome these limitations. Even then, the adoption would also need to be driven by commercial forces, and some of the limitations in terms of resilience, and the ability to offer multiple routes to a destination, may prove unattractive or difficult to overcome for the data-networking industry.

Notes and further reading

AES, SC-06-02, AES-X95 Draft proposal, open.

Device Bay (1997) Compaq Corporation, Intel Corporation and Microsoft Corporation. http://www.device-bay.org.

IP1394 Working Group, Internet Engineering Task Force. *IPv4 over IEEE 1394*, draft-ietf-ip1394-ipv4–18.

Keio University. 2-15-45 Mita, Minato-ku, Tokyo 108-8345, Japan. http://www.startap.net/igrid/jap-usa2.html

Leviton Telecom, 2222 – 222nd St, SE Bothell, WA 98021-4422, USA. http://www.levitontelcom.com/.

PAVO, Inc. 95 Yesler Way, Seattle, WA 98104, USA. http://www.pavo.com

Universal Serial Bus (1998a) *Device Class Specification for Audio Devices.* Release 1.0. http://www.usb.org.

Universal Serial Bus (1998b) *Device Class Specification for Audio Data Formats.* Release 1.0.

Universal Serial Bus (1998c) *Device Class Definition for Audio Devices.* Release 1.0.

Universal Serial Bus (1999) *Device Class Definition for MIDI Devices.* Release 1.0.

Universal Serial Bus Specification (1998a) *1.1 Motivation.* Revision 1.1. Compaq Computer Corporation, Intel Corporation, Microsoft Corporation, NEC Corporation.

Universal Serial Bus Specification (1998b) *1.2 Objectives.* Revision 1.1.

Universal Serial Bus Specification (1998c) *5.10 Special Considerations for Isochronous Transfers.* Revision 1.1.

1394 Trade Association Office. 1394 Trade Association, Regency Plaza, Suite 350, 2350 Mission College Boulevard, Santa Clara, CA 95054-1552, USA. http://www.1394ta.org.

7 Physical layer interfaces

7.1 Introduction

As stated in Chapter 1, there is no technology that can deliver unlimited network bandwidth on demand for all the possible types of service. 1394 and USB, discussed in the previous chapter, are capable of offering the QoS demanded by the functionality described as live performance in the section regarding QoS in Chapter 2, to a limited number of audio channels. The exact efficiency of the interface depends upon the implementation of the upper layer protocols, or in other words, how the audio data are loaded into the frame, and how transmission is managed.

As discussed, the most significant differentiation of network types is that made between circuit switched and multiple access networks. That is, from the view of the functionality of the rest of the communications model, the connection at the physical layer is a dedicated electrical circuit rather than being shared amongst multiple devices.

From another point of view, the traditional routing and patching of analogue audio signals create well-defined electrical circuits which are, as mentioned, network like, with circuit switched characteristics.

In data network circuit switching, networks may be temporary in nature, or may be set up within a central switch as a permanent connection. Telecommunications jargon, rather than data jargon, begins to infiltrate the terminology as a connection, or call, may be a permanent connection, or take just a few minutes or much less. Since the route which a particular call takes may be reassigned after use (or multiplexed with other calls), it can be described as virtual. Different technologies are able to efficiently create, hold open, and close calls. Those technologies that can open and close calls most quickly and efficiently are able to emulate shared media networks, and make efficient use of the available data rate.

This chapter concentrates on Fibre Channel and ATM. These two technologies are similar in that the significant feature of the communications policy is to manage the physical layer. Fibre Channel is also similar to IEEE 1394, in as much as both are described as possible solutions to the same kind of problems (fast and flexible device level communications), although Fibre Channel does not suffer from the early distance limitations of 1394. The solution applies to the problems of I/O bottleneck, and the bandwidth bottleneck, depending upon the outlook.

The solution requires a fast communication interface to move data between components within a particular system. Both IEEE 1394 and Fibre Channel are able to offer high data rates, capable of sustained and guaranteed loads, making them attractive for use within cluster environments and environments requiring a guaranteed QoS.

ATM on the other hand is a relatively young technology which has been compared to Fibre Channel because of the similarity with which the two operate. Under further inspection, it can be seen that ATM and Fibre Channel solve different parts of the same puzzle.

7.2 Cluster networks

Clustering takes many different forms, each form offering different advantages. A cluster system is a term coined to describe either a 'parallel or distributed system consisting of a collection of interconnected whole computers to become a single, unified computing resource' (Pfister, 1995). The general characteristics for the different types of cluster networks are:

- High-availability clustering, which allows the workload of a failed system to transfer to another node within the cluster (fail-over).
- Administrative clustering, which involves clustering systems to simplify administration. Resources are allocated and managed across the cluster, but each application still runs on one device or node.

- Application clustering, which provides management of a specific application across a cluster through tight integration with its own application programming interface – for example, a post-production software suite.
- Scalability clustering, which spreads a specific workload across multiple nodes with the use of system functions (Turner, 1999).

A system designed to meet one of the descriptions above does not exclude the possibility of the system being capable of meeting criteria in one of the other classifications, although in general cost will be the determining factor.

High-availability clusters are used in situations where the system is required to run for long periods of time without incurring downtime, as described in Chapter 1.

Administrative clustering includes redundant resources for restarting applications.

Scalability clustering is designed to increase the capacity and growth potential of a system, thereby reducing the amount of time that the system is out of action for upgrades and maintenance.

7.2.1 Applications for clustering

Cluster technology is still in the early stages of deployment and new initiatives appear constantly. One such initiative is the virtual interface (VI) architecture specification that attempts to define an industry-standard architecture for communication within clusters of servers and workstations.

Most early cluster implementations were installed within the same facility, because communication interfaces were not available to consider geographical distance. As such, common methods of communication use fibre media over short distances between a number of computers, normally servers, that are not distributed and contain all the required components of a computing system. In some cases, clustering is performed within a single device, containing multiple CPUs, and this is more synonymous with parallel computing environments, for consideration when enterprise level performance is required from a server. With the advent of Fibre Channel, clustering can be performed between campus buildings or over greater distances.

System area networks
One area of research into applications for a transparent communications technology capable of transmitting over long distances is within distributed systems. A distributed system is one where

the physical locations of the system components (CPU, disk etc.) do not relate to one another. In other words, the components normally housed within a system unit are instead located in various positions around a network.

The network encompassing the components of the system is known as the system area network. Subjectively, a system area network may encompass a logical or virtual part of a wider network, and be spread over various geographical distances, from a few centimetres, to metres or kilometres. More objectively, a system area network is the interface between the system components.

In general, the purpose of a cluster is to share computing load over several systems whilst operating as a single cohesive system. A component in the system might be classified as hardware or software, depending upon what defines the system in a particular instance. If a component fails, a successful cluster will continue to provide the service, despite being impaired. Additionally, if more processing power is needed then new components can be plugged in, and the performance of the system as a whole improves (*Windows NT Clustering Architecture White Paper*).

Consideration should be given to the use of clusters in environments where high availability and performance of the computer system are key measurements for the success of the system.

Storage area networks

Apart from increasing the distances between directly communicating devices, one of the most prevalent applications for cluster technology is within storage area networks (SAN). It should be noted that the acronym SAN can be used to mean storage or system area networks. Since system area networks have already been covered, the acronym SAN will hereafter mean storage area network.

SANs are networks working at the back-end of a user community, hidden away and used for the purpose of transmitting storage and I/O generated data.

Using Fibre Channel (FC) mappings in layer FC-4, communication between peripherals is supported for SCSI and serial storage architecture (SSA from IBM) for instance. This means that all storage-related information, such as data retrieved as a result of a system request, or data moved from one storage device to another, e.g. during a backup, will not effect the performance of the user network in any way. It is perceived that data housed in one place is easier to manage because storage systems can copy data for testing, routine backup, and transfer between databases, without burdening the hosts they serve. Operationally, this indicates a requirement for a controlled environment in which to house the storage, such as a machine room or data centre.

Storage area networks can be implemented as centralized or decentralized installations. A centralized SAN is described when multiple host processors use a single central storage system, such as a RAID disk array (Figure 7.1). Each system may rely entirely on the storage system to store all data, although current implementations still require a boot device, which is usually disk storage. The centralized storage topology is commonly employed to cluster systems together for purposes of redundancy in case of failure. In decentralized configurations, each server can be assigned its own disk arrays, meaning that several arrays may exist (see Figure 7.2).

Figure 7.1 Centralized SAN configuration in which multiple host processors utilize a single disk commodity. In this example, machines are assigned disks only within the array which they are attached to.

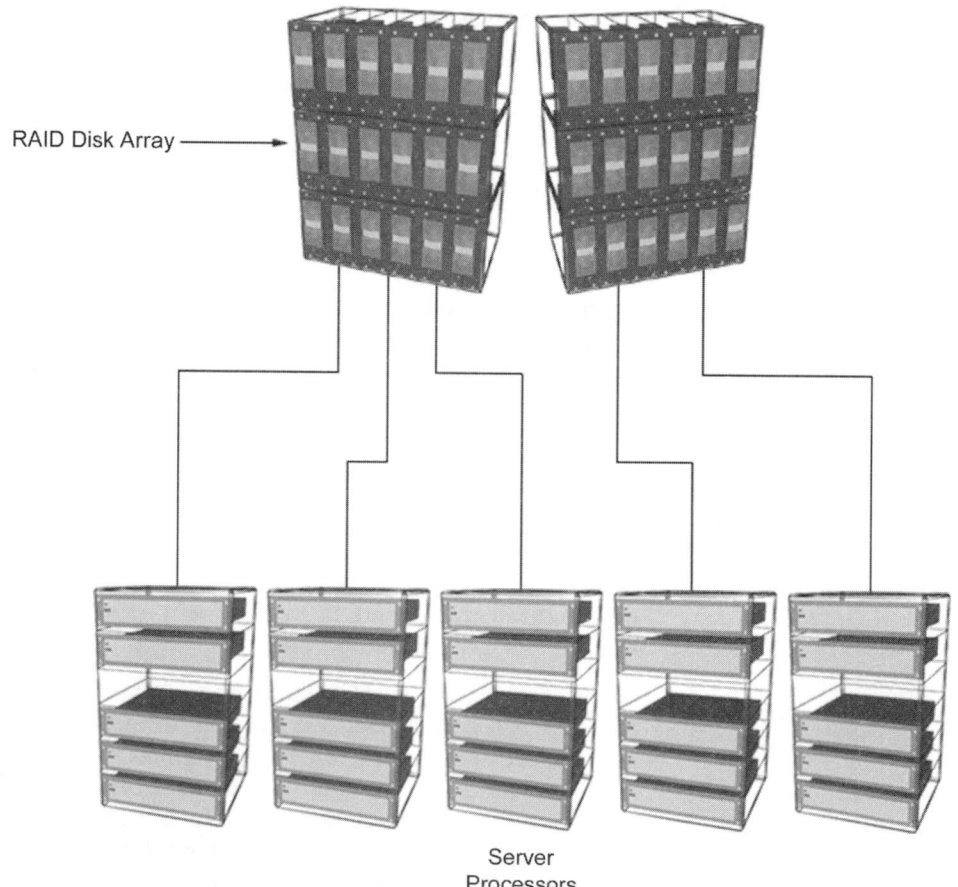

RAID Disk Array

Server
Processors

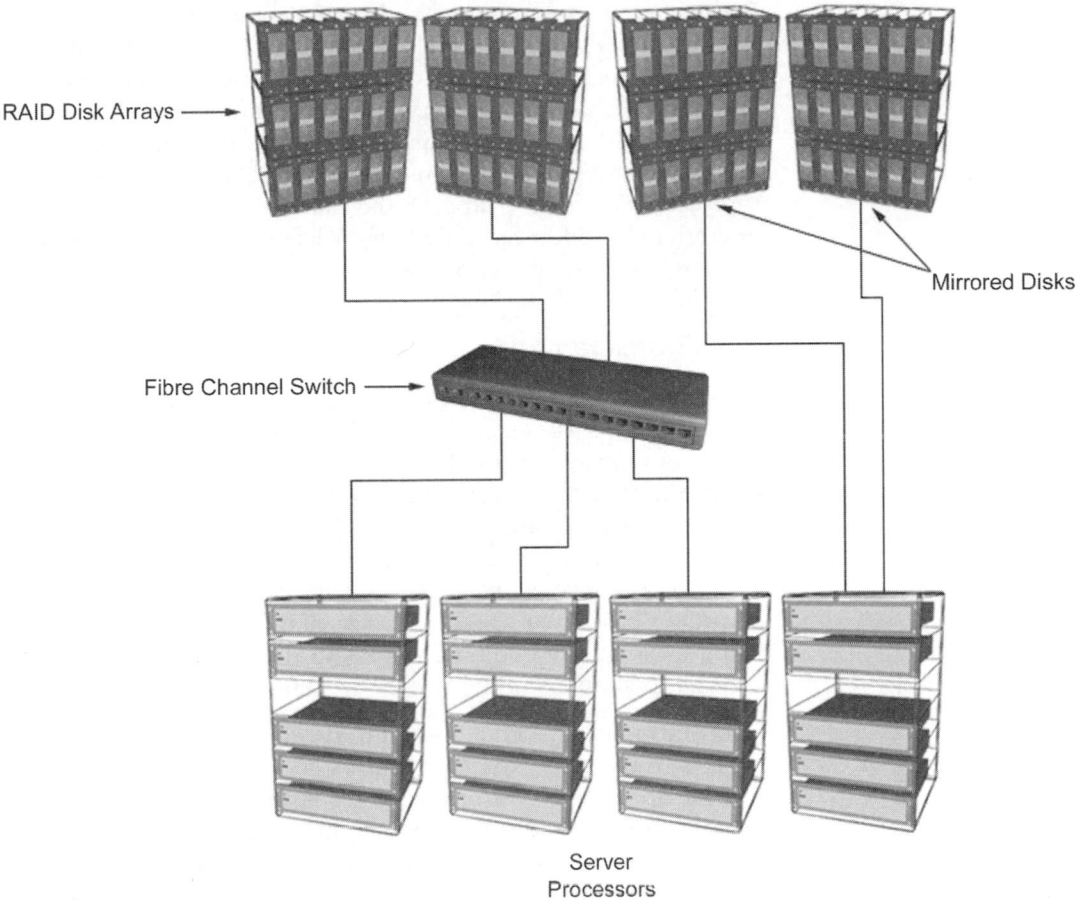

RAID Disk Arrays

Mirrored Disks

Fibre Channel Switch

Server
Processors

Figure 7.2 A decentralized storage area network, in which multiple processors each utilize a dedicated storage facility. In this example, more hosts are attached through a switch or hub, allowing the disk resource to be shared further. This is typically the upgrade path from Figure 7.1.

FC is the prevalent interface technology for storage area networks, thereby allowing distance to be considered flexible. Furthermore, FC support for existing peripheral and network interfaces allows for easy integration with existing systems, making FC a driving force during the early stages of cluster deployment. In such an installation, the hosts communicate with the storage system via SCSI, as well as communicating with each other via IP over the same physical FC topology.

7.2.2 Applications for SANs within media industries

SAN architectures have made early appearances within the media industries. The benefit of installing such a system is for the archival and general storage of large amounts of data, because of the flexible way in which storage can be upgraded.

Because of the huge amount of data generated by commercial media enterprises, additional benefits within products aimed at this sector include the ability to swap storage in and out with ease. This means that a complete project can be stored on disks, and removed from the chassis, in much the same way that analogue (or digital) tape is removed from the tape machine and replaced with a new tape at the start of a new project. This operational support has only been feasible since the cost of hard disks fell to compete with tape.

7.2.3 Considerations

Storage of digital audio and visual data occupies a relatively large amount of disk space, depending upon how long the media plays for and what resolution is used, when compared to other applications such as word processors and so on. As a result of this, server class computing devices installed for audio and video use are of little use unless enough storage is available to store complete projects. At the same time, in order to reap the benefits of project-based work, where more than one operator may be using a file at any one time, access to the disk needs to be fast enough to support uninterrupted working for those operators.

In order to install such a system to meet these requirements, all aspects of the storage system and its communicating links need to be inspected to ensure the best possible throughput of data from the disk to the user. This includes the speed at which the disk can retrieve data, and a good indication of this is the measurement of seek-time of the disk, which can be obtained from generally available benchmarking software tools. The seek-time depends upon the efficiency of the disk at finding a particular location, along with the speed at which it spins, normally given in revolutions per minute.

7.3 Fibre Channel

Fibre Channel is the result of an initiative investigating a transport mechanism capable of carrying the data away from the I/O system of a component, including disks, CPU class devices, and others. The applied interface supports a faster and more distributed generation of hardware systems.

7.3.1 Introduction

Although the conceptual work initially assumed a fibre physical transport medium, it was realized during development that large parts of the interface were not reliant on fibre, and so copper support was added later (Zoltán Meggyesi – see Notes and further reading).

7.3.2 General description

Current Fibre Channel implementations supply data rates from 133 Mbits/s up to 1062 Mbits/s, over link distances of 10 km or less. The types of transmission media which can support the interface are specifically identified within the standards. The standards allow expansion of the data rate up to 4 Gbits/s, with the limitations currently within the performance of applied examples.

The link is considered to be duplex in nature, although transmission and receive signals are sent through separate physical media, so it is actually half-duplex. The confusion arises because there are two complete fibre connections, one for transmission, and one for receive (TX/RX). The physical link behaves in this way, even if the connection is made over copper.

Co-ordination of the link is performed by the exchange. The port starting the exchange is called the originator and the answering port is known as the responder. A set of related frames flowing in one direction is called a sequence, whereas exchange describes a transaction or operation of related sequences occurring in both directions. Data are sent in frames that can contain a variable payload, from zero to 2148 bytes long. An adjustable header increases the frame up to the minimum specified size of 64 bytes, where necessary, and a 24-byte header is the mandatory minimum header size. Each frame is marked by a 4-byte start of frame sequence. A 4-byte checksum (32-bit) field is located at the end of the frame, shown in Figure 7.3.

The sliding window technique, as the flexible header size is known, is used to control the flow of data, and can be used to support provision of a deterministic QoS.

7.3.3 Standards and administration

FC was arrived at within the hierarchy of the ANSI approved National Committee for Information Technology Standards (NCITS, commonly pronounced 'Insights') which is a commercially focused standards-making organization (NCITS – see Notes and further reading).

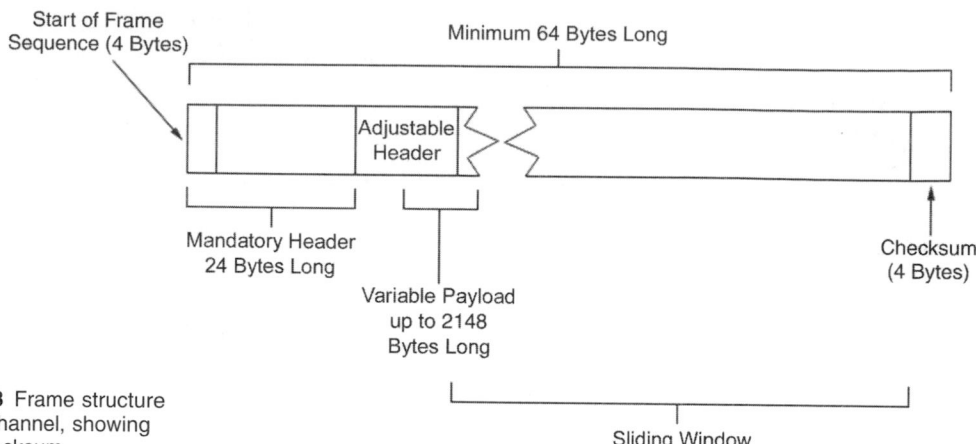

Figure 7.3 Frame structure in Fibre Channel, showing 4-byte checksum.

From 1961 to 1996, NCITS operated under the name Accredited Standards Committee X3, Information. References to X3 within products imply the administration of the originating committee in reference to compliance with other products and standards. The NCITS mission statement is '. . . to represent commercial interests by consensus of opinion in the matter of multimedia (MPEG/JPEG), intercommunication among computing devices and information systems, . . . storage media, . . . and database[s], . . .' (copyright 2000 NCITS, http://www.ncits.org).

The Accredited Standards Committee X3T11 has the specific title of Technical Committee for Device Level Interfaces and is the primary committee responsible for Fibre Channel, and others.

Technical Committee (TC) T11 is the parent committee of task groups T11.1, T11.2, T11.3, and T11.4, each defining its own particular subject within the standard. The parent co-ordinates the work of all the task groups and retains overall responsibility for work in the area of device level interfaces. T11 assumed the programme of work of the predecessor Task Group XT9.3 following a reorganization of TC X3T9. T11 held its first meeting in February 1994.

T11 is specifically responsible for standards development in the areas of intelligent peripheral interface (IPI), high performance parallel interface (HIPPI) and Fibre Channel (FC). The standardization initiative began in the mid-1980s, with the primary focus of T11 being the Fibre Channel (FC) family of standards.

One of the most interesting aspects of the FC standards family is the provision for mappings which allow the management of the physical layer to act upon information contained within upper layer protocols. The original set of mappings includes IP, which allows the technology to operate as a network solution for LAN-type networks and in any other situation where IP has been specified. The task group T11.4 was responsible for all FC projects defining protocol mappings and was dissolved on 7 October 1999. The corresponding European technical committee to T11 is ISO/IEC JTC 1 SC25/WG4.

The Fibre Channel Systems Initiative was formed in 1992 by three companies interested in defining SCSI and IP interoperability within the context of existing Fibre Channel standards (Fibre Channel Industry Association – see Notes and further reading). The companies Hewlett Packard, IBM, and Sun Microsystems developed a series of profiles for the Fibre Channel technology, whereupon the role of supporting and continuing development was handed over to the Fibre Channel Association (Jain, 1996). The companies retain an interest in Fibre Channel technologies, with products available from the respective portfolios.

7.3.4 Topologies

An unusual feature of Fibre Channel is the variety of physical connection topologies that can be configured. Point to point connection, ring (or loop in Fibre Channel terminology), or connection through a central switch are all valid topologies.

Point to point

The most basic configuration is as a point to point connection achieved by connecting two devices together. Each interface on the device is called an N_Port and the fibre connectors are shown in Figure 7.4. All N_Ports support point to point

Figure 7.4 Fibre Channel fibre connector.

connection but additionally, some N_Ports support multiple devices connected in a loop.

Devices may be provided with more than one N_Port, and each N_Port can support a unique point to point connection, thereby allowing the attachment of multiple devices, as shown in Figure 7.5a. For example, if a disk array was supplied with more than one N_Port, the disk array could act as a software repository for one computer system attached to each N_Port.

Figure 7.5 (a) Different connection topologies possible in Fibre Channel point to point configuration. Notice that each device may only support as many connections as there are available connectors. When devices communicate through another device (as between device A and B shown), other devices act as pass-through or repeater devices.

Point to point connections act in the same way as when connected in a loop and so may be considered as a two-node loop.

Arbitrated loop

When devices are connected in a ring, the resulting topology is known as Fibre Channel arbitrated loop (FC-AL). FC-AL supports up to 127 devices connected in a series, continuing back around to the originator, as shown in Figure 7.5b. Attachment ports supporting connection in a loop are known as NL_Ports.

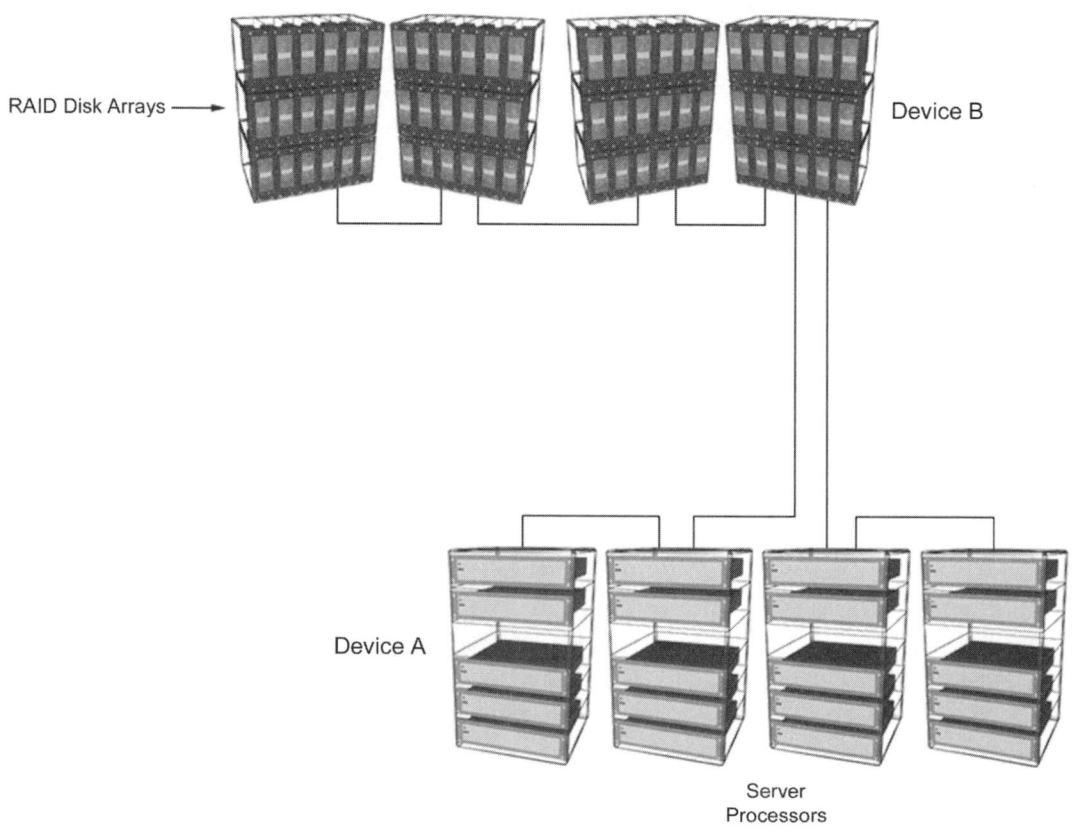

RAID Disk Arrays

Device B

Device A

Server
Processors

Following arbitration, the conversation between two devices A and C (from Figure 7.5b) is established. When the conversation takes place, it is considered to be a point to point connection (or two-node loop), with all intersecting nodes acting as line repeaters. Once connection between two ports is established, no other ports can communicate on the loop until the current connection is brought to an end. Once the conversation is closed, the loop is made available for arbitration, and a new connection may be established.

Because of the nature of connection within FC-AL, bringing one node down will cause the entire loop to fail. In order to prevent this, an electronic arrangement known as a port by-pass circuit is used. This device allows a node to be electronically removed from the loop, without interrupting traffic or data integrity on the interface.

FC-AL can be cabled through a concentrator, which is a hub-like device specifically for connecting FC in a loop, and these devices most often contain the electrical switch to heal

Figure 7.5 (b) Point to point topology in Fibre Channel, shown connected in a loop to create a Fibre Channel arbitrated loop. The operating principle is identicle to that in Figure 7.5(a). In such a configuration, devices would be cabled to a hub, in a similar physical arrangement to token ring, and as with token ring, each FC-AL hub is equipped with ring in and ring out ports.

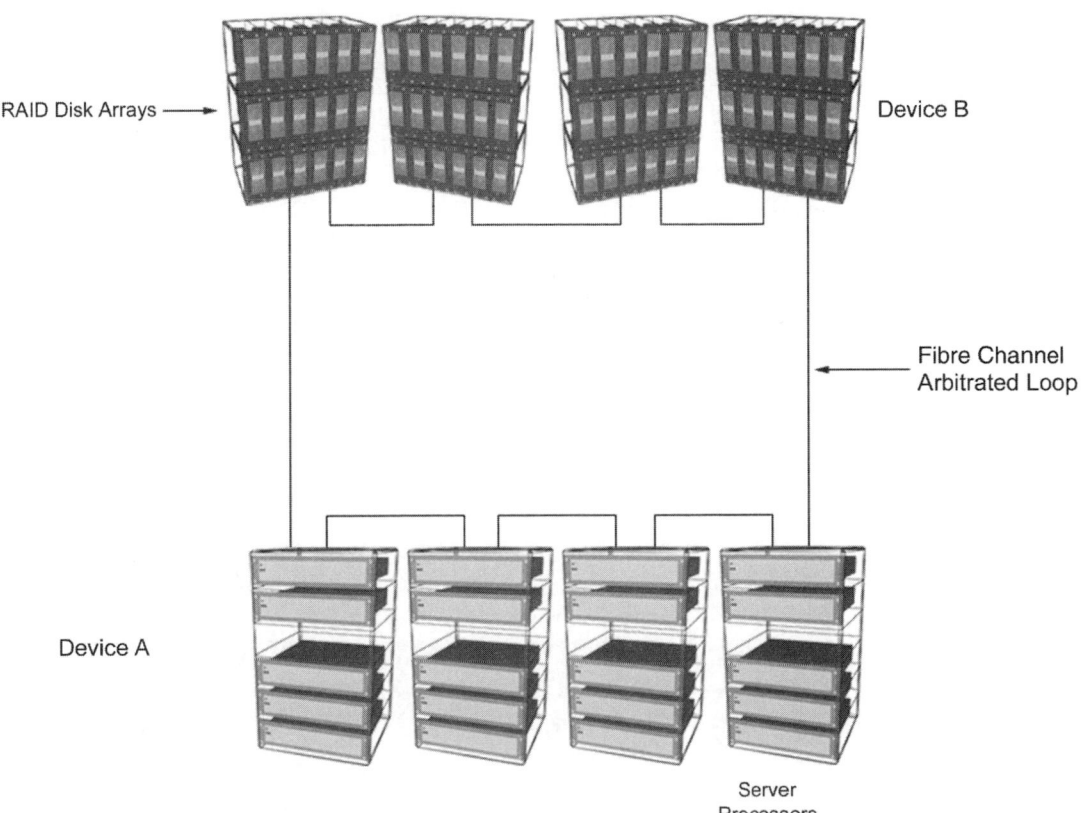

RAID Disk Arrays

Device B

Fibre Channel
Arbitrated Loop

Device A

Server
Processors

the loop in the event of node failure, by-passing the non-operational node to transmit between the nearest operational nodes.

Switch or fabric

A Fibre Channel switch, or a network of connected switches, is called a fabric because of the implied topology, shown in Figure 7.6. The switch appears to be somewhat like the network hub described in Chapter 3, as it acts as a concentrator for cables and connections in the same way as a network hub or telecommunications node. The connection interfaces appearing on the front of the switch are known as F_Ports. F_Ports are specifically to be found on the concentrator, and can be connected to N_Ports or NL_Ports at the other end of the link.

Figure 7.6 Mesh or fabric topology as connected with Fibre Channel F_Ports, along with the schematic representation, illustrating how the term 'fabric' is arrived at.

In fabric configuration, a physical layer connection is created inside a central switch, directly between communicating devices. All that is required of the switch or fabric is to manage the multiple point to point connections.

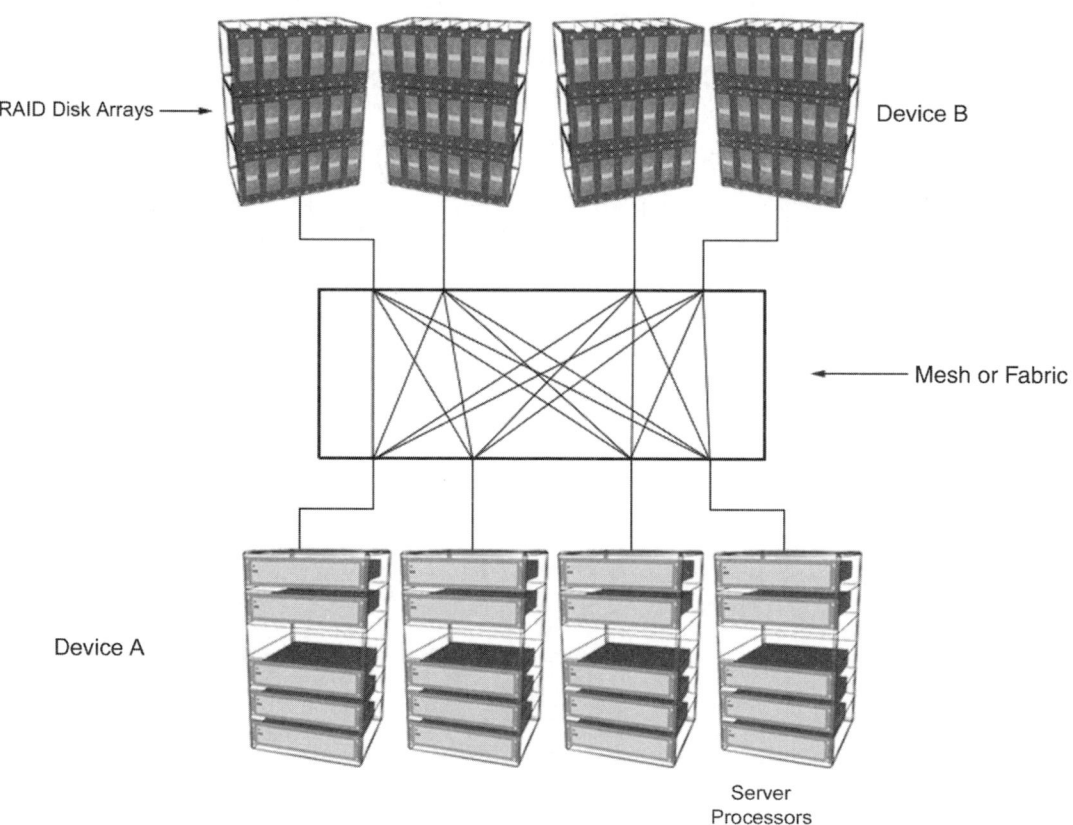

RAID Disk Arrays

Device B

Mesh or Fabric

Device A

Server
Processors

7.3.5 Quality of service

The physical layer connection is known as a channel, meaning that FC is connection oriented, providing a direct or switched point to point connection for dedicated use. Circuit switched networks are typically hardware-intensive, with the switch managing thousands or millions of connections. The limit is set on the capability of the switch, as a function of the speed of the switch itself. This performance figure is calculated as being the slowest speed of either the silicon switch itself, or the backplane through which connections are made. The switch mechanism is generally capable of switching rapidly enough to manage multiple connections, and can open and close connections in a manner satisfactory to bursty network traffic.

By creating a dedicated circuit for communication, the FC central switch maintains a deterministic data rate upon which the application may rely in order to provide a level of service that has applications in real-time transmission applications, such as audio.

FC switch products provide connection data rates in the order of 1 Gbit/s, and the speeds of the switch continue to improve beyond 38 full-time connections.

7.3.6 Communication model

Conceptualized in Figure 7.7, FC is structured as a set of hierarchical functions supporting the management of the various topological configurations that have been discussed.

The transmission media is defined within layer FC-0, which describes fibre optic and electrical parameters for a variety of data rates. Fibre Channel's position within X3 illustrates the scrutiny of the management of the physical layer, peering with IPI (intelligent peripheral interface) and HIPPI (high performance peripheral interface).

As such, the real work lies within abstracted layers that manage the physical link, allowing the varied topology, deterministic QoS and low error rate, as shown in Figure 7.8. In simple terms, the interface can adopt several different personalities, allowing it to be transparent to the devices using it.

7.3.7 FC-0 physical definitions

FC-0 is designed for the use of flexible technologies to meet a wide range of interface requirements. An end-to-end communication link may consist of different technologies to achieve the

Figure 7.7 Fibre Channel communications model is structured as a set of hierarchical functions supporting the management of topology configuration.

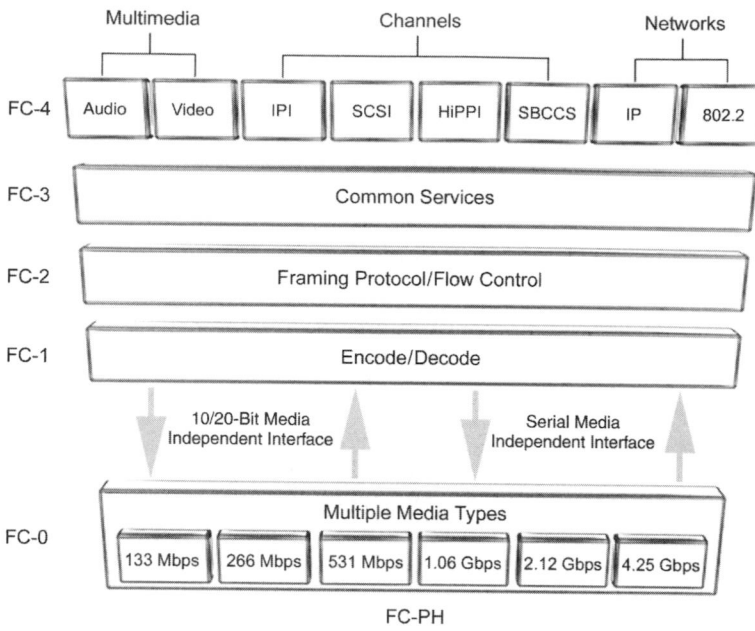

Figure 7.8 Fibre Channel abstracted position within the 7-layer model. This shows the support for connection regimes behind the technology. The same principle applies to ATM.

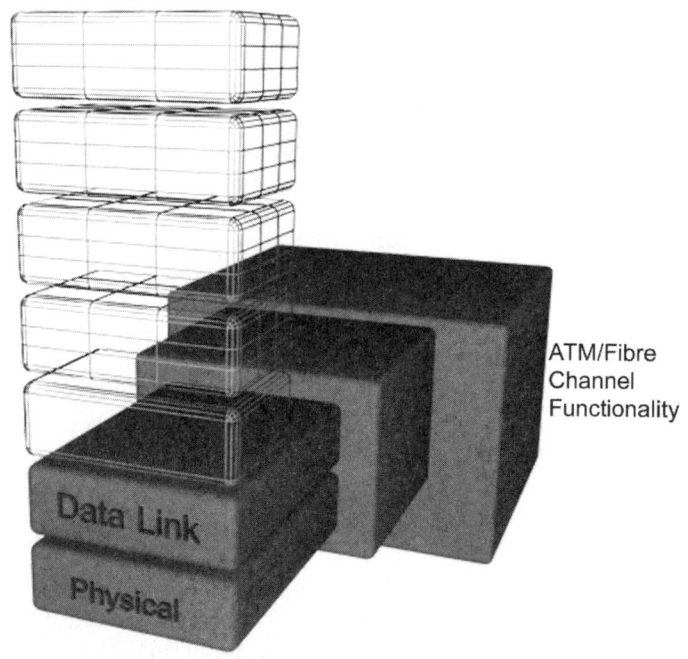

maximum performance within suitable criteria. Figure 7.9 shows how the endpoint device may be connected by a copper link to the fabric, whereas interswitch connections or loop connections can be one or other form of optical media.

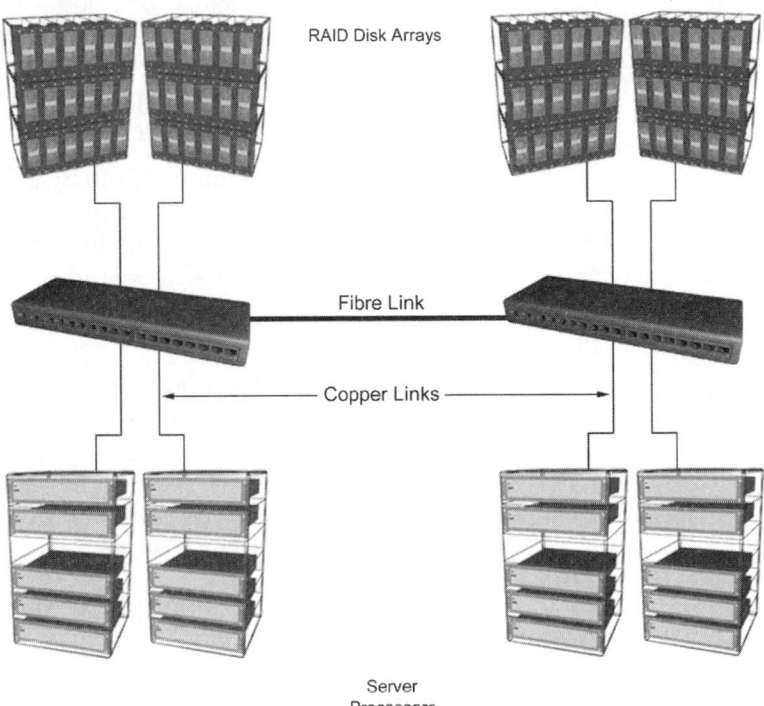

Figure 7.9 Copper and fibre connections in Fibre Channel topologies. This flexible arrangement allows disks to be assigned to any processor from any part of the disk commodity.

FC-0 specifies a safety system for optical media called the open fibre control system (OFC) which allows the transmitting laser to exceed the power ratings defined by laser safety standards. OFC is invoked when an open fibre condition occurs in the link, or in other words, when a fibre is disconnected. The device detecting the open condition pulses the transmission laser at a low duty cycle within recommended safety parameters. Detecting the pulsing signal from the other end of the link, the originating device also transmits a pulse at a low power cycle. When the open path is restored, both ports receive the pulsing signals, and after a handshaking procedure, the connection is restored.

Cables
Fibre Channel supports the following distances for electrical media types: Coax/twinax support FC running at a data rate of 1.0625 Gbits/s over 24 m, or 266 Mbits/s over 47 m.

For fibre-based transmission media, 9 μm single mode long-wave laser transmission media supports data rates of 1.0625 Gbits/s over distances up to 10 km, whilst the 50 μm short-wave laser defined for multimode transmission systems supports data rates up to 1.0625 Gbits/s over 300 m, or 266 Mbits/s over 2 km. Finally, 62.5 μm multimode long-wave LED implementations support a data rate of 266 Mbits/s over distances up to 1 km or 132 Mbits/s up to distances of 500 m. It should be noted that these distance figures represent the distance between nodes, rather than the total distance around the loop.

In summary, although multimode cable supports a higher bandwidth than single mode fibre, the dissipation caused by light bouncing around inside the thicker multimode cable means that multimode is less useful over long distances.

Connectors

In order to facilitate the interchangeable operation of upper layer protocols over a variety of physical media, FC defines four physical connection types. These are split into fibre optic connectors and electrical connectors. Fibre optic media is supported through the definition of multimode and single mode media and connectors (Figure 7.4).

Multimode and single mode connectors are generally differentiated through the use of a simple keying device, such as a slot or guiding pin located on one surface of the connector. It should be noted that the differentiation is not standardized in any way and caution should be taken when working in a mixed environment. However, multimode fibre is thicker in appearance than single mode fibre.

Copper cables are supported by the use of a standard STP 9-pin connector, also known as DB9 connectors, as shown in Figure 7.10. In order to differentiate the connector, the five centre holes are filled in, and the four remaining holes are allocated as transmit and receive pairs, as shown in Figure 7.10.

7.3.8 FC-1 encoding

Layer FC-1 defines encoding and decoding of data onto the media, special characters, and error control. Binary data are encoded onto the media using a form of NRZ, dependent on the media type.

The special or transmission character is created by encoding 8-bit bytes into a 10-bit string, and calling it a character. The

Figure 7.10 Fibre Channel DB9 connector for copper-based STP links showing missing pins.

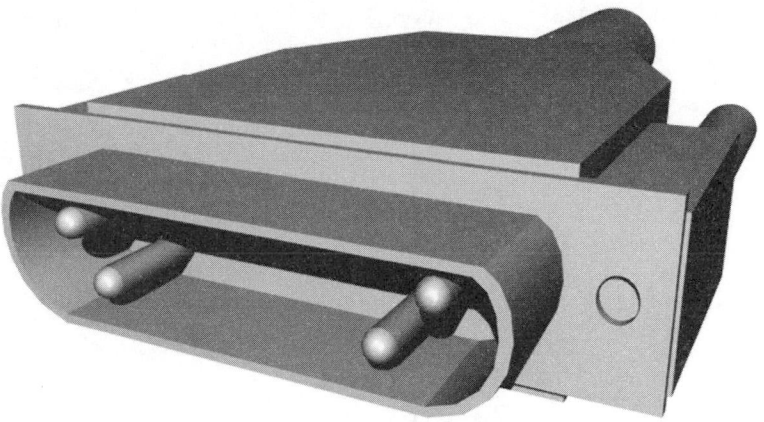

mechanism that creates the additional bits is an insertion of two error checking bits into the data, and is discussed in the next section.

Error control

The particular mechanism for error control is known as 8b/10b and was originally proposed for adoption within the standard by IBM for its robustness when transmitted over long distances. IBM also use the technique in their ESCOM interface.

8b/10b was proposed by IBM from whence it came, and is more generally known as disparity.

Disparity can be explained by the representation of three states, positive, negative, and neutral, for instance. The three states represent whether there are more 1s or more 0s in the data word or whether there are the same quantities of each. In this case, a negative disparity indicates more 1s in the character, and conversely a positive disparity indicates more 0s, leaving a neutral disparity to indicate when the quantities are the same. The disparity bits are placed in the first 6 bits and the last 4 bits create the transmission character, as shown in Figure 7.11.

A transmission character can represent one of two transmission codes. Data are placed into D class characters and protocol management data are represented in K class special characters. Identification is by means of placement within a frame, and is part of the synchronization function, which is described overleaf.

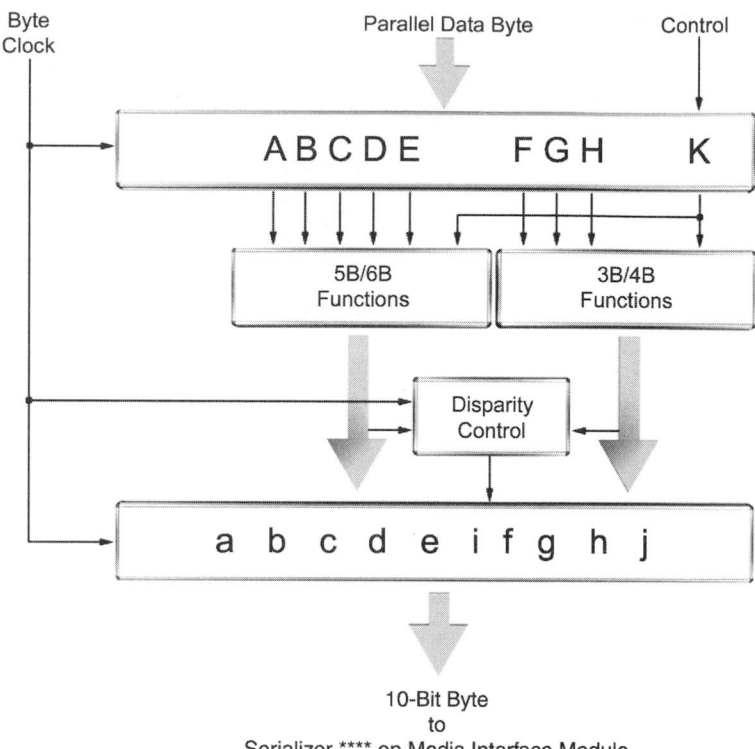

Figure 7.11 Disparity implementation within Fibre Channel.

Upon receipt of the character, the receiving device calculates its own disparity from the character, and compares it with the one inside the character. If the results do not match, the receiver indicates a violation condition. Codes detected at the receiver which are not D or K class also result in violation errors. At this layer, no action for error recovery is defined, which is left to behaviour mappings in FC-4.

Synchronization

The exact representation of disparity has the effect that long strings of 0s will not appear in the transmission. This is achieved by inserting a 1 to indicate there are more 0s in the string. This is put to good use to enable arbitrary clock recovery by ensuring that a minimum number of transitions are present in the serial bit stream.

Transmission word

A transmission word is composed of four transmission characters in a frame-like sequenced set arrangement. A word can either be a data type or an ordered set. The distinction is made within the first transmission character. If the first transmission character is a data byte then all of the other characters are also data characters and so the set contains data.

If the first transmission character is a K character then the word is an ordered set. Other transmission characters within the set are data characters. Ordered sets are covered in more detail under FC-2

7.3.9 FC-2 the framing protocol

FC-2 is the signalling or framing protocol and defines the frame structure, sequencing, and the mechanisms for controlling service classes. To aid in the transport of data across the link, the following building blocks are defined by the standard. This layer defines the necessary criteria to transmit and receive data, starting with the definitions of ordered sets.

Ordered set

An ordered set contains a special control function with one of three possible meanings. The first use is as a frame delimiter, which defines the class of service that is required, and contains start of frame and end of frame delimiters. Multiple SOF and EOF delimiters are defined for the Fabric and N_Port sequence control.

The second use is called a primitive sequence, and is three identical ordered sets transmitted consecutively. In this way, control signals can be sent within the stream of data. The control codes are used for control of the link, for instance to automatically recover following a failure. Control sequences are transmitted and repeated continuously to indicate the specific status of the interface. When a primitive sequence is received and recognized, a corresponding idle packet is transmitted in response.

Lastly, an ordered set can be a primitive signal. Primitive signals come in two types, the first of which is used for flow control related to the buffers of the devices attached to the link, and the second is for idling.

As may be deduced from this description, Fibre Channel actively transmits across the media, in an idling state, even when there are no data to send, in order to maintain bit and byte synchronization.

Frames

Before being broken up into transmission characters, it is the FC-2 layer's responsibility to group data into basic units for transfer, more classically known as frames, and to reassemble data from the frames at the other end of the transmission media. As we have seen, frames can be broadly categorized into data frames and link control frames. Each frame contains the payload, source and destination address, as well as link control information, and 32-bit CRC.

Data frames may be used as link or device data frames and link control frames are classified as acknowledge (ACK) and response frames. Link response frames include instructions for when the link is busy or to indicate an error state.

Each frame begins and ends with a 4-byte delimiter (as shown in Figure 7.12) and the frame header immediately follows the start-of-frame delimiter. The header is used to control link applications, device protocol transfers, and to detect missing frames or frames received in the incorrect order. An optional header may contain further link control information. The payload is carried in a 2112-byte long field. The 32-bit CRC precedes the EOF delimiter.

Figure 7.12 Fibre Channel frame structure, showing detail of the 4-byte start-of-frame and end-of-frame delimiters, along with the expanded detail for the frame header.

Sequence

A sequence is formed by a set of one or more related frames transmitted in one direction from one N_Port to another. Each frame within a sequence is uniquely numbered with a sequence counter by FC-2 for tracking purposes.

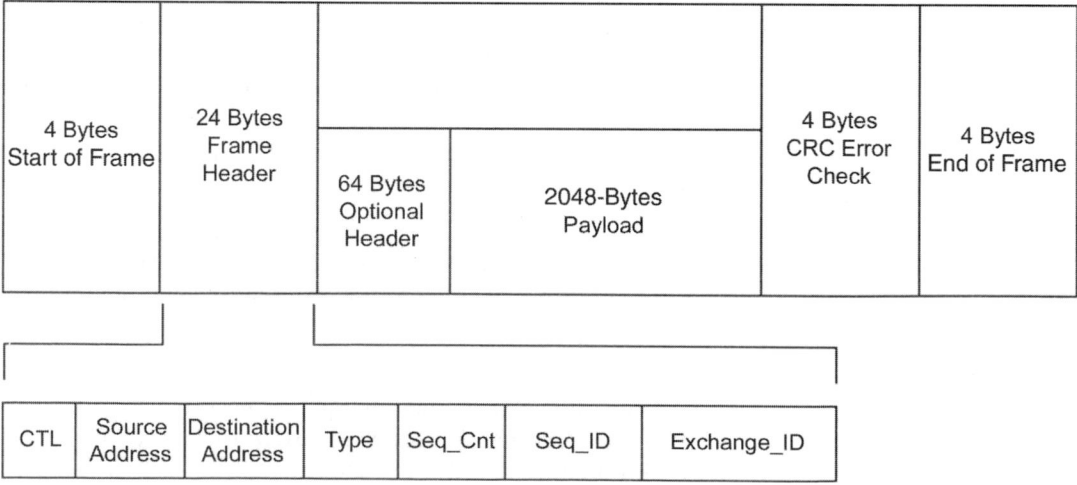

Exchange

Exchanges are the largest construct understood by FC-2. An exchange is made up of multiple sequences, transmitted in one or both directions between two N_Ports. Only one sequence may be active at any one time for a given exchange. To send data in the opposite direction, sequence initiative is passed from one port to another.

Frame structure

The frame structure is illustrated in Figure 7.12, in which the header is expanded for detail.

Protocol

FC protocols describe mechanisms for arbitrating transmission. For instance, primitive sequence protocols are based on primitive sequences and describe the activity that takes place in the event of a failure of the link.

When a port is first attached to the fabric, the fabric login protocol allows the communication of parameters from the N_Port to the fabric. At the other end of the link, the N_Port login protocol communicates parameters with another N_Port before data are transmitted.

The data transfer protocol is used to transfer data and describes the methods of transferring upper layer protocol data using the flow control management facility. Lastly, the N_Port logout protocol frees unused resources in a controlled way and controls closure of the link.

Flow control

Both the initiator and recipient manage control of the data flow through a system of credit. Credit is based on the number of buffers, or the amount of memory, allocated to a port. A counter counts the number of data frames that have not been acknowledged by the recipient and action is taken accordingly. The management of the flow between N_Ports or between N_Port and fabric is associated with the service class.

Class 3 uses only buffer-to-buffer flow control. Class 2 frames use both types of flow control. Class 1 service uses end-to-end control to pace the flow of frames between N_Ports.

When end-to-end flow control is used, the recipient is responsible for acknowledging the valid data frames by use of an acknowledge frame. When the number of receive buffers in the device is insufficient to store the incoming frame, the recipient sends a busy frame. If an error is encountered, a reject frame is sent to the initiator. The flow control technique used within service class 2 is a combination of the techniques used for class 1 and class 3. Class 3 uses buffer-to-buffer flow control managed between N_Port and F_Port or between N_Ports in point to point topology. Each port is responsible for managing the buffer-to-buffer credit counter, which is established during the login procedure. This simple mechanism requires the recipient port to send a receiver ready signal to the transmitting port, which indicates that there are free memory buffers in which to place incoming frames.

Service classes

Service classes categorize different types of traffic, so that they can be handled in different ways. The service class can be selected based on the characteristics of the application to which the interface is to be put. Considerations include such parameters as packet length, transmission duration, and services allocated by the fabric login protocol.

Class 1 is a service providing dedicated connections, like a permanent circuit. Once established, a class 1 connection is retained and guaranteed by the Fabric. This service guarantees the maximum bandwidth between two N_Ports and so is best suited to sustained high throughput transactions such as streaming. Frames are always delivered to the destination port in the same order as they are transmitted.

Class 2 is a packet switched, connectionless service allowing bandwidth to be shared by multiplexing frames from multiple sources onto the same channel or channels. The fabric does not guarantee the order of the delivery of frames, which may therefore be delivered out of order. This service can be used when the connection set-up time is greater than the latency of a short message. If delivery cannot be made due to congestion, a busy frame is returned and the sender tries again. Both classes 1 and 2 send acknowledgement frames to confirm delivery.

Class 3 service is identical to class 2, except that frame delivery is not confirmed. Known as a datagram switching service, class 3 provides the most efficient transmission by removing confirmation overhead, and might be likened to offering a QoS similar to UDP in relation to IP. Class 3 service might be considered for real-time applications, where information not received in timely fashion is valueless.

An optional service mode called intermix is also described by the FC standards. Intermix is a variation of class 1 service, where frames are guaranteed at a portion of the data rate, but class 2 and class 3 frames are multiplexed onto the channel when sufficient data rate is available without impairing the class 1 channel. Such a service might be useful to guarantee service for a particular purpose on a multi-user network, whilst retaining the ability for users of the network to continue operation. This configuration might impede the normal network users, since their bandwidth has less priority over the required service, if the network load grew above the threshold set by the permanent circuit.

7.3.10 FC-3 layer

FC-3 is the first of the layers that has no understanding of the topology of the installation, which is otherwise handled by FC-1 and FC-2. However, FC-3 does respond to the topology and current status of the installation and the capabilities of the ports that are immediately attached. For instance, FC-3 understands if multiple ports are attached, and whether the ports are capable of participating in multiport operations such as multicasting.

The FC-3 layer provides common services for advanced features such as multicasting and striping.

Striping

In the context of FC, the term striping is used to describe the parallel transmission of data through multiple N_Ports in the fashion of a parallel interface. This technique is used to achieve a higher data rate by allowing multiple links to simultaneously transmit a single information unit across multiple N_Ports in a parallel fashion, but can only be used in instances where the device supports multiple interfaces.

Hunt groups

FC-3 also provides the ability for more than one port to respond to the same alias address. N_Ports are placed together in hunt groups by assigning a number to the N_Port. The number represents a virtual N_Port, to which the interface can transmit. In this way, any number of N_Ports can be assigned to a hunt group and multicasting is achieved.

Multicasting

The ability to broadcast a message to groups of devices on the network is also handled here, and is known as the multicast service. FC multicast can be likened to a network broadcast, as it delivers a single transmission to multiple destination ports. The combination of hunt groups and multicasting offers a choice between sending data to one, all, or a defined subset of devices on the network.

7.3.11 FC-4 layer

FC-4 is the last level in the conceptual structure, and achieves the goal of the Fibre Channel initiative in combining aspects of networks and switches, by allowing protocol addressing to affect a switch mechanism. This allows the switch to take on a particular operating behaviour according to which upper layer protocols contain the data.

Mappings

FC-4 achieves the objective by defining rules to allow upper layer protocols to access functionality defined within the FC standard. Known as mappings, these might be thought of as a means of translation or as an interface to the functionality of FC for upper layer protocols. For instance, support for MADI channels could conceivably be configured as a means to bandwidth appropriation. Through FC-4, multiple protocol types can be supported concurrently over the same physical interface.

The following protocols are specifically defined: small computer system interface (SCSI), intelligent peripheral interface (IPI), high performance parallel interface (HIPPI), framing protocol, Internet protocol (IP), link encapsulation (FC-LE), single byte command code set mapping (SBCCS) and IEEE 802.2.

The rules specifically allow for transformation of information units into Fibre Channel sequences and exchanges, and back again.

Upper layer protocols

Upper layer protocols (ULP) allow two devices to communicate. For example, when a CPU sends data to a disk, it is the ULP which provides the general transparency to I/O, which allows the configuration of geographically displaced devices, mentioned earlier. ULP exists for each protocol that Fibre Channel can transport.

7.3.12 Fibre Channel arbitrated loop (FC-AL)

Arbitrated loop can be configured in a number of geographically different ways. Token ring and modern Ethernet, are generally associated with a star cabling topology, whereas FC can be cabled in the manner of a loop or a star. Cabling in a loop is considered a small to medium installation, as it avoids the cost of a hub and additional cabling. A single loop, enclosed in this way, is known as a private loop. The use of a hub, on the other hand, also allows the connection to other hubs, or to attach to a fabric, as shown in the various topological-related figures.

Each N_Port independently discovers when it is attached in a loop and addresses are assigned upon initialization. There is no loop master since devices in the loop communicate with each other in point to point fashion, with intersecting devices acting as line repeaters for the duration of the conversation.

7.3.13 Fibre Channel and asynchronous transfer mode (ATM)

The concept behind Fibre Channel (FC) is administration of the physical layer, as shown in Figure 7.8, making it comparable with ATM, which is also a circuit switching technology with a similar set of specifications.

By mapping the ATM adaptation layer for computer data (AAL5, discussed in more detail in the following sections) into FC-4, Fibre Channel maps ATM functions over its physical layer. Therefore, Fibre Channel is compatible with ATM and can co-exist with it.

7.3.14 Applications

Fibre Channel's goal of providing a transparent and ubiquitous transport mechanism inspired the development of applications for extending the communication distance between processor and storage (disk).

Processor and storage are generally housed within the chassis of the same device, although communication between CPU and storage still needs to occur for the data to be moved and placed onto the storage device. This form of communication uses one or other communication buses built onto the motherboard of the device, such as the PCI bus, SCSI, or IDE. These buses are housed in the controlled environment inside the device and tend to be parallel in nature, spanning short distances. For instance, modern versions of SCSI are still limited to distances of a few metres in the same order as IEEE 1394. By replacing the SCSI cable and using FC-4 mapping for SCSI, the disk can be removed from the device and placed in a different building, several miles away. Theoretically, the practice can be extended to separate any physical component from within a device, meaning that the device can now be architecturally located over geographical distances. This concept is expanded in the section on system area networks.

Early implementations within the media and multimedia industries are within storage area networks, which are a form of cluster network and are covered earlier in this chapter.

7.4 Asynchronous transfer mode

7.4.1 Introduction

Direct comparisons have been made between ATM and Fibre Channel in the form of reports and white papers, and these often include Gigabit Ethernet in the discussion. The three are placed together for comparison in this way because of the similar data rates that each can provide. However, direct comparisons of this type are difficult to assess since the different technologies are designed for different purposes.

Removing Gigabit Ethernet from the discussion for the purposes of this chapter, Fibre Channel and ATM have quite a lot in common. Both incorporate methods for topology discovery for instance and generally break down the tasks necessary for management of a connection, so that they can be described in various ways. In this way, both have achieved flexible interfaces, capable of offering speeds in excess of normal networks by the use of switch mechanisms to provide virtual circuits between communicating devices. Furthermore, ATM is made to appear as a hybrid technology, able to offer network-like functionality through LAN emulation service (LANE). This is a little like the function of FC-4 layer in providing direct access to ATM functionality from upper layer protocols.

7.4.2 General description

ATM is a pure switching technology, allowing the transfer of multiple types of data including conventional computer data, audio, image, and video whilst maintaining different QoS levels for different communication channels.

ATM offers scalable data rates for local area networks and interconnection with telecommunications type services. ATM scales from LAN data rates, starting at 25 Mbits/s, up to telecommunications networks supporting 622 Mbits/s and beyond. ATM is not limited to any particular cable type or medium, as provision is made within the standards framework to continue adding new media types.

The framework has led to the provision of a flexible array of specifications throughout the communications model. The comprehensive list of specifications is not reproduced in full here, since this subject area has already filled a number of excellent books on the subject (Gadecki and Heckart, 1997). Of more interest to the streaming media community is ATM's ability to maintain a deterministic QoS over a circuit, even when network traffic is bursty. Packet switching and circuit switching are combined by allocating bandwidth on demand, efficiently and cost effectively, whilst at the same time guaranteeing bandwidth for delay sensitive applications such as audio and video.

ATM works by breaking up variable length data delivered from (for instance) another network, into small, equally sized parcels called cells. A cell is 53 bytes, and consists of a 5-byte header and a 48-byte data payload. The cells are transmitted on the ATM network and the original data in the packet are reassembled at the destination by concatenating the cell payloads.

To reduce processing overheads that otherwise cause delay, the packet header information does not have to be repeated in each ATM cell. It appears only in the first cell of a packet transmission sequence being mapped or absorbed into the ATM payload, just like the rest of the packet.

ATM is connection oriented and communications are established by setting up a virtual channel between the sender and receivers, much like circuit switching. The set-up process establishes the route from source to destination, and so complex routing information does not need to be repeated in each cell.

In essence, this system is similar to setting up a normal telephone call. The address or telephone number is sent through the network to establish the connection. Information is then sent, or the conversation takes place, over the pre-established path through the network, making the ATM connection virtual in nature. Two types of virtual connection exist. The first type is a permanent virtual circuit (PVC), which is defined by the network manager or operator responsible for configuring the ATM switch. The second type of virtual connection is a switched virtual circuit (SVC), which is connected and disconnected after the two end stations have finished their conversation, just like a telephone call.

The signalling protocol may also support the setting up of a point to multi-point connection, where one station transmits to more than one other station. As with Fibre Channel, this is called a multicast, and an example of this would be video transmission from a service such as video on demand.

In the absence of enough formal standards, many multi-vendor ATM networks are manually configured using PVCs. The interim switching signalling protocol (PNNI) defines the signalling mechanisms between ATM switches and is key to multi-vendor support for SVCs. The full standard is expected to support dynamic routing and the exchange of addresses and link state information between switches.

7.4.3 Standards and administration

ATM is an enormous undertaking, and is not likely to be finished for some time. Early conceptualization and development occurred with the telecommunications industry, and the initiative was based on the ISDN standards covered in the next chapter. As such, ATM was intended as a wide area network protocol.

The ATM Forum (see Notes and further reading) was formed in 1991 as an international non-profit making organization with the objective of accelerating the use of ATM products and services through convergence of interoperability specifications. The ATM Forum Technical Committee works with other worldwide standards bodies selecting appropriate standards, resolving differences between standards, and recommending new standards when existing ones are absent or inappropriate. The Technical Committee was created as a single worldwide committee in order to promote a single set of specifications to ensure interoperability between different vendor's products. The Technical Committee consists of several working groups, which investigate different areas of ATM technology.

The ATM Forum launched a major strategic initiative called the Anchorage Accord, intended to guarantee a stable platform for network implementations. In addition to designating a set of foundations and expanded feature specifications, it also establishes criteria to ensure interoperability of ATM products and services between current and future specifications.

The Anchorage Accord comprises 60 specifications and outlines foundation specifications needed to build an ATM infrastructure. It also identifies expanded feature specifications which are needed to enable migration to ATM multi-service networks. The Accord establishes criteria to ensure interoperability of ATM products and services between current and future specifications. The specifications within the Anchorage Accord are available for download at the ATM Forum Internet site (*The Anchorage Accord* ftp://ftp.atmforum.com/pub) (Figure 7.13).

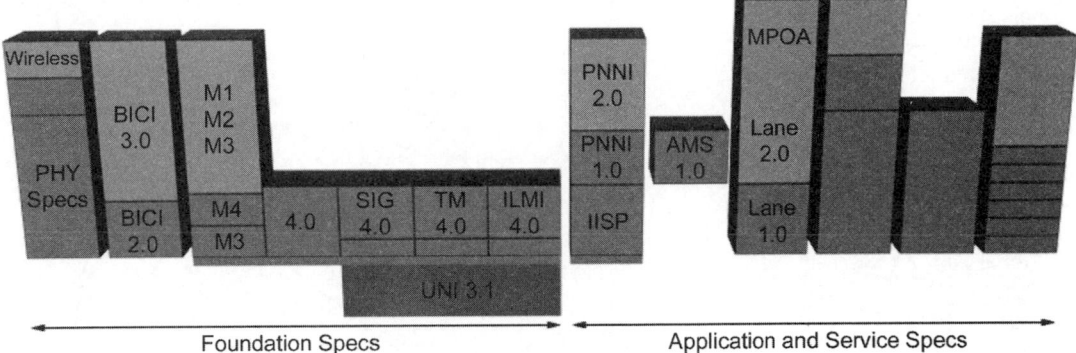

Foundation Specs   Application and Service Specs

Figure 7.13 Outline of the Anchorage Accord model, developed for ATM specifications.

Notable amongst the documents contained within the Accord are the five documents pertaining to LAN emulation, and a further document entitled 'AudioVisual Multimedia Services: Video on Demand Specification v1.1'.

7.4.4 Topologies

When installing an ATM network, it is common to avoid the expense of individual host adapters for every device requiring attachment to the network. Instead, direct ATM connections are chosen for those devices that warrant the additional data rate, such as servers, but supply an edge switch connection for other devices. An edge switch is a network device that supplies classic LAN connections, such as Ethernet or fast Ethernet, via standard RJ45 ports. A number of ports are made available on such edge switches, which are normally housed within the vicinity of a cabling closet.

The ATM diagrammatic representation of network components is shown in Figure 7.14.

A simple network is shown in Figure 7.15 and is cabled as shown. Linking three switches in this way provides a redundant link in case of failure. A switch can be connected to a number of other switches in a manner which allows for very loose connection strategies and still provides redundant links, as shown. If a switch is dual homed (two connections to a single outbound device), then the redundant link is used for load balancing, and is therefore active, rather than truly redundant. In case of failure of one or other of the links, all of the traffic is placed on the other link.

An ATM network is referred to as a cloud, rather than a fabric, and the apparently random configuration is managed by ATM to provide a resilient and intelligent communications interface.

Switch

Router

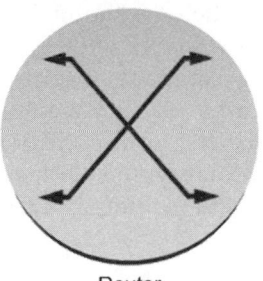

Figure 7.14 Schematic representation of ATM network components.

231

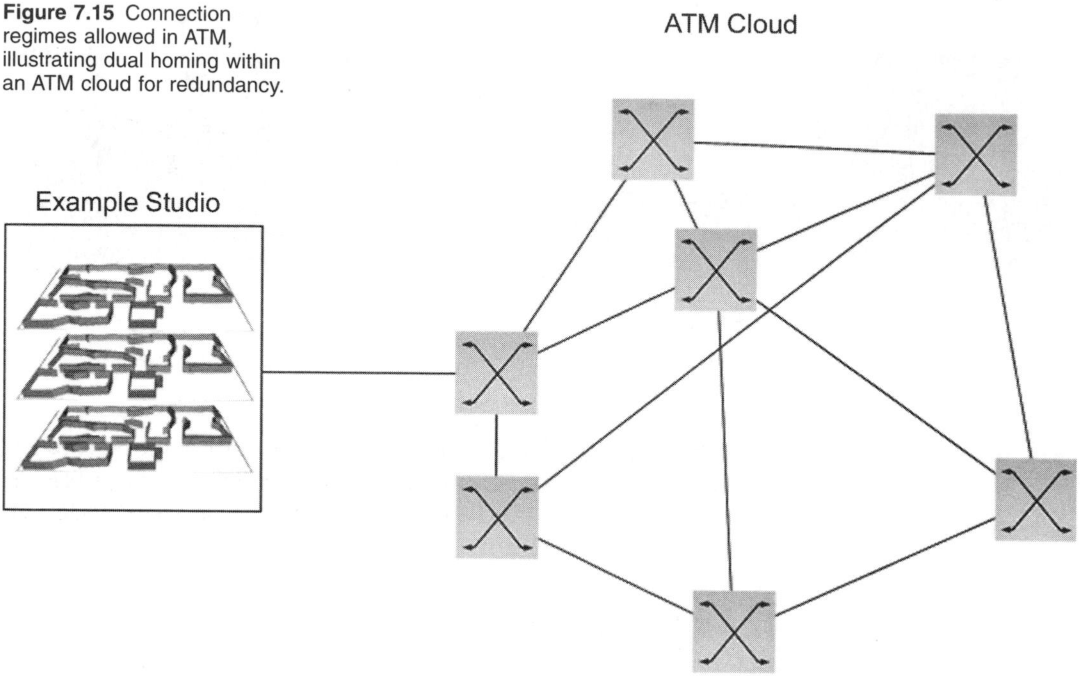

Figure 7.15 Connection regimes allowed in ATM, illustrating dual homing within an ATM cloud for redundancy.

ATM Cloud

Example Studio

7.4.5 Quality of service

The ATM cell size allows a deterministic delay across the network, since each component within the ATM network deals with one whole cell at a time. It is early days for ATM but much research is being carried out in areas such as QoS management and multimedia applications (Vogel, 1995).

7.4.6 Communication model

The theoretical model of how ATM traffic is handled is set out in layers similar to the OSI model laid down by the ISO. Figure 7.16 shows a detailed breakdown of the four ATM layers with their names. Control and management planes accompany the user plane: user plane describes how the transfer of information between users takes place through the layers; control plane defines the signalling through the layers; management plane provides control of the ATM switch or node and includes overall layer management and plane management.

Physical layer

As ever, the physical layer contains definitions for the physical interfaces, media and so on, as well as framing protocols. Many

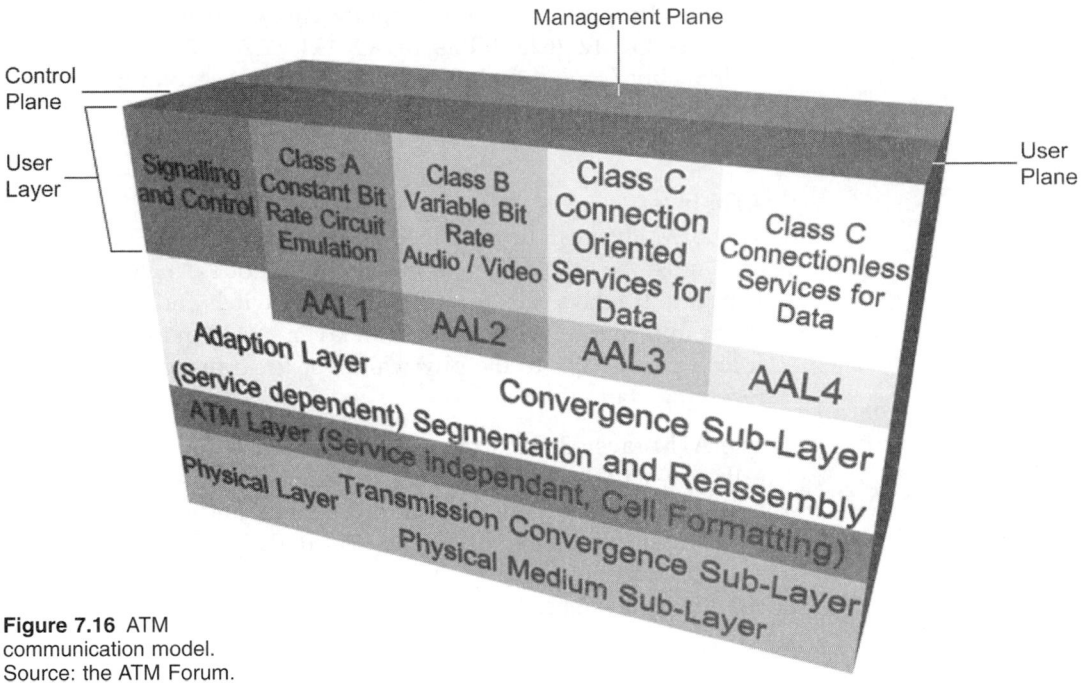

Figure 7.16 ATM
communication model.
Source: the ATM Forum.

proposals have been put forward, each able to carry a 53-byte cell but over different cable types and at different speeds. In theory, an ATM cell could traverse several different types of physical layer by moving between different private ATM domains without difficulty.

There are five main ATM forum specifications providing connectivity between devices. These are:

- DS-3 (or T3) access at 45 Mbits/s based on coaxial cable and intended as a public ATM user network interface (UNI) in the US.
- 100 Mbits/s local ATM multimode fibre interface, based on the FDDI standard was defined to allow the early deployment of low-cost private UNIs.
- 155 Mbits/s fibre for private ATM, specifies use of the Fibre Channel standard 8b/10b encoding rules on multimode fibre.
- 15 Mbits/s STP is the STP cable version of the 155 Mbits/s fibre standard.
- STS3c is based on the 155 Mbits/s SONET (synchronous optical network – see section on telecommunications), offering a choice of monomode fibre for public UNI and multimode for private UNI.

ATM is also planned to run at optical carrier speeds above OC-3, such as OC-12 (622 Mbits/s), OC-24 (1.244 Gbits/s), OC-48 (2.488 Gbits/s), and higher, such as used on telecommunications backbones.

ATM layer

The ATM layer can be described as the guts of the ATM standard and must be implemented in all ATM nodes in a switching network. It relays cells through an ATM switch and passes them to the appropriate ATM adaptation layer (AAL) in the next node. It also passes cells to the physical layer for transportation to the receiving station.

The ATM layer defines the cell structure and the way in which cells are actually passed across the logical connections: header creation and checking; cell multiplexing; and demuxing (the undoing of multiplexing). The logical paths are based on virtual path connections (VPC) or collections of VPCs known as virtual channel collections.

Figure 7.17 shows the structure of the five header bytes of a 53-byte UNI cell. Each header includes a path indicator comprising the virtual path identifier (VPI) and virtual channel identifier (VCI). These elements determine the route which the cell is to follow through the switches, each of which then maintains a table of active connections. The format of the cell has been agreed, but opinions differ over matters of traffic management.

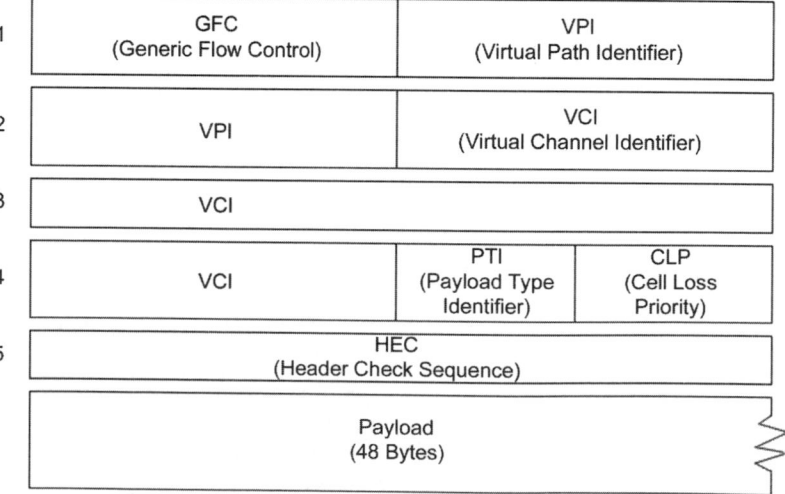

Figure 7.17 ATM cell structure.

ATM adaptation layer

The ATM adaptation layer is responsible for providing the interface between the user application and ATM. It performs the segmentation and reassembly of cells, taking variable data packets and converting them into 53-byte cells and vice versa.

7.4.7 Service classes

Four classes of adaptation are defined, supporting five AAL service types. AAL-5 combines AAL-3 and AAL-4. AAL types are:

- AAL-1: emulation of dedicated circuits for voice.
- AAL-2: audio and video support.
- AAL-3: connection-oriented data.
- AAL-4: connectionless data.
- AAL-5 connectionless/connection-oriented data without a 4-byte header in the data payload.

Extensions to the AAL layer define common data protocols such as TCP/IP.

Figure 7.18 shows the relationship between the AALs and the four classes of service, plus the characteristics of each service class. Class A (AAL-1) traffic is a constant bit rate and connection oriented, such as encoded voice traffic and video, with a direct

Figure 7.18 Relationship between ATM adaptation layers and the four available classes of service.

	Class A	Class B	Class C	Class D
Timing Sensitivity Between Connection	Yes		No	
Bit Rate	Constant	Variable		
Connection Service	Connection Oriented			Connectionless
AAL Types that Apply	1	2	3 / 4,5	3 / 4

timing relationship between traffic source and destination. Class B (AAL-2) traffic is a variable bit rate and connection oriented, such as variable bit rate (or compressed) voice and video traffic (the most common form of voice and video traffic). The same timing issues apply as in class A but the traffic may be sent in bursts to the AAL.

Data are sent to AAL-2 from higher level protocols at fixed intervals but in variable packet sizes, which are segmented by the AAL into fixed size cells and transferred to the ATM layer. The AAL performs the reverse function when receiving call streams from the ATM layer.

Class C (AAL-3/4) traffic is a bursty, variable bit rate and connection oriented, such as TCP/IP and most other shared access protocols. Class C service is made up of the convergence sub-layer and a segmentation and reassembly sub-layer. The convergence sub-layer maps higher level protocols, using both connectionless and connection-oriented data streams, into the AAL layer, and passes it to the segmentation and reassembly sub-layer, mapping the traffic into the ATM layer for submission.

AAL-5 was created as an efficient alternative to AAL-3/4 and also supports class C service. It is designed to support data-only messages and is simple to implement, allowing only one packet to be transmitted at a time, compared with AAL-3/4 which allows multiple packets to be interleaved on a virtual connection.

Class D (AAL-3/4) traffic is connectionless with a variable bit rate traffic. It is a service which runs on top of AAL-3/4 and such traffic is not sensitive to source and destination timing. Class D does not need to establish a specific connection to the destination. Instead, the sending node sends data to the server, which establishes a separate connection with the destination without the sending station being aware of the connection. Class D is bandwidth on demand as the connection is established when the transmission is started. As such, it is beneficial to low-volume, limited connectivity applications such as terminal support. Its major drawback is a long set-up time, making it unsuitable for time sensitive applications.

7.4.8 Segmentation and reassembly

The ATM standard specifies how to transfer messages that are longer than the ATM cell size, such as LAN traffic. To transfer a block of data, a header and trailer are added to the information to form a packet. To transmit a packet, it is segmented into pieces small enough to fit into an ATM cell. The cells are then transmitted over the ATM link. When no information is sent over the ATM link, the ATM transmitters send idle cells. At the receiving end, the idle cells are removed and data cells are reassembled into the original packet.

The order of ATM cells is maintained over the ATM link. Error correction and retransmissions are not part of the ATM standard, and so are left to upper layer protocol layers in the end station and lower layer physical transmission standards. ATM cells belonging to different packets may be interleaved over the transmission line. Different data from different users may therefore be transmitted simultaneously over the link. This ensures fair access and enables multimedia applications, such as the simultaneous transmission of video and audio as part of a videophone call.

Applications

Although large parts of the telephone infrastructure are running on ATM, for more general use, the applications of ATM remain rare. The standard set, based on the Anchorage Accord, is constantly changing and being added to. The work of the ATM Forum consists of investigating other communications standards, to glean possible areas of compatibility. This is why the user layer is not covered in the above descriptions.

One notable example of a successful application occurred within SohoNet (see Notes and further reading), in which film and video production facilities purchased ATM switches from the same manufacturer. By doing so, the organizations were able to connect the switches together to create a high speed network used for the purposes of transferring film and audio through different stages of production. The network now extends beyond its roots in Soho, London, to connect via a private Internet link to facilities in the United States.

The reader interested in practical ATM implementations within the media industry is guided to Rooholamini and Cherkassky (2000), Nahrstedt and Smith (2000), and The Internet 2 project, as well as numerous research works, publications and sources on the Internet.

Notes and further reading

Gadecki, Cathy and Heckart, Christine (1997) *ATM For Dummies*®. IDG Books Worldwide, Inc.

Fibre Channel Industry Association. 404 Balboa Street, San Fransisco, CA 94118, USA. http://www.fibrechannel.com.

Jain, Raj (1996) *Gigabit Networking Standards: Fibre Channel and HIPPI.* The Ohio State University, Columbus, OH 43210-1277, USA.

Nahrstedt, Klara and Smith, Jonathan M. (2000) *The QOS Broker,* Vol. 2, No. 1. IEEE Multimedia.

NCITS, Information Technology Industry Council. 1250 Eye Street, NW/Suite 200, Washington, DC 20005, USA. http://www.ncits.org.

Pfister, G. (1995) *In Search of Clusters.* Prentice-Hall, Inc., page 72.

The ATM Forum (1997) 2570 West El Camino Real, Suite 304, Mountain View, CA 94040-1313, USA. http://www.atmforum.com.

The Internet2 project, UCAID. 1112 16th Street, NW Washington, DC 20036, USA. Considers aspects of quality of service, network structure, usage models, and Ipv6 and can be found at http://www.internet2.org.

Rooholamini, Reza and Cherkassky, Vladimir (2000) *ATM-Based Multimedia Servers,* Vol. 2, No. 1. IEEE Multimedia.

Sohonet Ltd. 19–23 Wells Street, London W1V 3FP.

Turner, Vernon (1999) *Compaq AlphaServers Step Up to IT Center Consolidation.* An IDC White Paper. 5 Speen Street, Framingham, MA 01701, USA. Compaq Computer Corporation. ttp://www.digital.com/alphaserver/news/idc_it_center_paper.html.

Vogel, Andreas (1995) *Distributed Multimedia and QOS: A Survey,* Vol. 2, No. 2. IEEE Multimedia.

Windows NT Clustering Architecture White Paper. Microsoft Corporation. One Microsoft Way, Redmond, Washington 98052-6399, USA. http://www.microsoft.com/ntserver/ntserverenterprise/techdetails/prodarch/clustarchit.asp.

Zoltán, Meggyesi. KFKI–RMKI Research Institute for Particle and Nuclear Physics, H-1525 Budapest, POB 49, Hungary. E-mail: zoltan.meggyesi@cern.ch.

8 Telecommunications networks

This chapter discusses geographically very large networks. From the previous chapter, Fibre Channel is suitable for delivering data over distances of up to 10 km. ATM functionality can be mapped into fibre channel so that ATM is transported over fibre channel links. In still larger networks, ATM is used as a mode of communication between nodes, where a node is best described as a switch that manages multiple connections.

The history of ATM within the telecommunications companies indicates its position as a backbone technology within telecommunications infrastructures. Descriptions of ATM do not complete the telecommunications picture, however, and this chapter presents some of the missing components relevant to data communications.

8.1 Introduction

The early phone network consisted of an analogue signalling system connecting telephone users directly via interconnecting wires. The term plain old telephone system (POTS) was coined in the US to describe the service of voice communication through a telephone handset and sometimes the infrastructure supporting that service. One important aspect of this service is the ability to use the telephone during an electrical failure.

The POTS infrastructure is inefficient and prone to breakdown and noise because of the age and variation of environments in which it is installed. As a result, it does not lend itself easily to the transmission of data, especially over long-distance connections. Packet based digital switching systems began replacing the old infrastructure in the 1960s, but due to their size and complexity, the final connection from the local central office to the customer equipment is mainly an analogue connection.

8.1.1 Character

In classifying networks by geographical distance, and arriving at telecommunication networks, further categorization by the service that is offered upon them is useful. For instance, voice, data, message, or image transmission, are all telecommunications services. Networks in this chapter may also be local or global, international, national, or within a municipal area, in scope. Different physical layer techniques are used, such as cable, radio frequencies, and microwave, depending upon the installation and specific purpose of the link.

In terms of data transmission, these connections serve as the interface between LANs and generally include the main backbone of telecommunication networks. The relationship between the types of network are shown in Figure 8.1 and the pattern created by the links might be likened to the delivery systems for electricity or water, transport networks such as road and rail, or even the blood vessels in the body.

Figure 8.1 shows how a close-up view of how one section might look in the neighbourhood of Example Studio, and the pattern of increasing complexity continues as the data filter out to encompass the LAN and the individual computer that is generating or receiving the data. It is possible to see the pattern of increasing complexity towards the edge of the network as the detail level is increased. This phenomenon is called edge-centric, and can be thought of as being fractal in nature. Interesting research on the structure of networks can also be found within the Internet2 project (Internet2 International – see Notes and further reading). The fractal nature of networks has also been explored at British Telecom (Appleby, 1994).

8.1.2 The line of demarcation

At the highest level of detail, the mapped data are moving within a private network belonging to Example Studio. The point where responsibility for the transmission of data passes between the

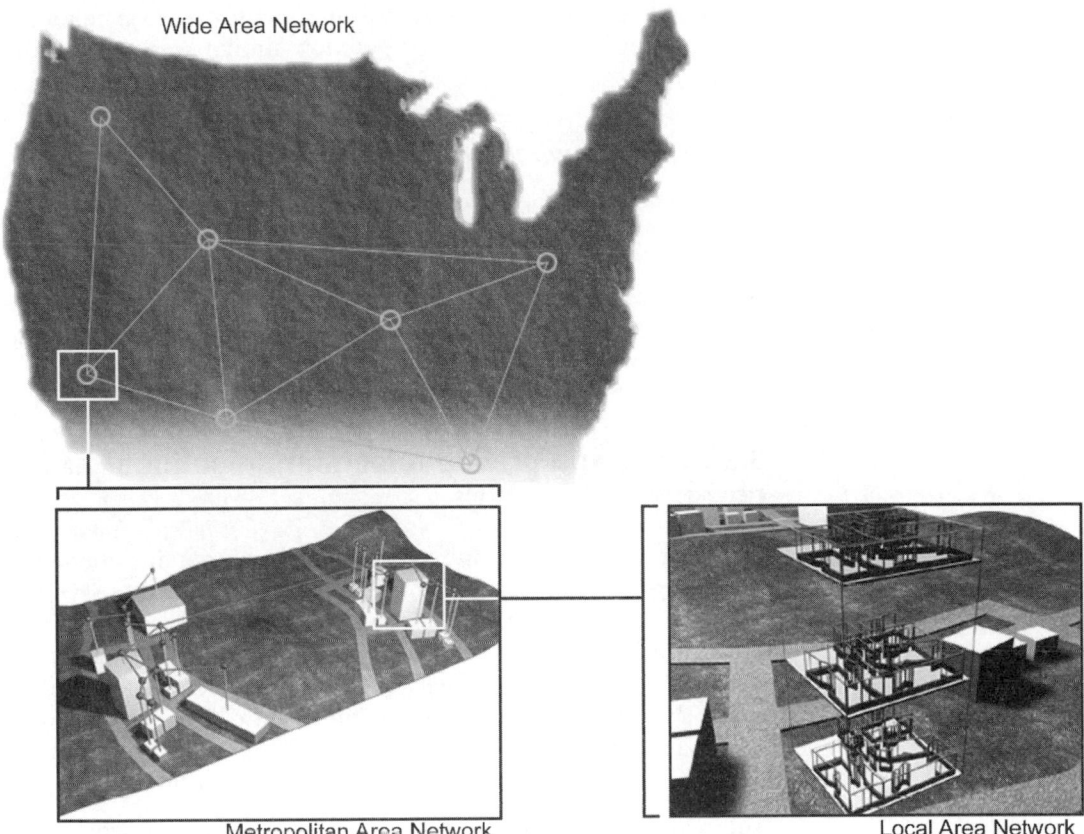

Wide Area Network

Metropolitan Area Network

Local Area Network

Figure 8.1 Relationship between local area networks, municipal area networks, and wide area networks. Although the example networks illustrated are imaginary, wide area networks can be more or less complex, and thousands may exist in a national infrastructure. The metropolitan area network (also municipal) connects urban areas. Local area networks focus on the premises with Example Studio shown. The point of connection between the LAN and the MAN (or WAN) is called the line of demarcation.

institution and the telecommunications provider is called the line of demarcation. The provision of services in this instance is normally documented within some agreement, which includes the service levels. Although the line of demarcation may be moved to unfamiliar locations, such as at the edge of the customers' property, the dependency is upon the exact nature of service that has been purchased.

A telecommunications provider can only make guarantees whilst the data are travelling on their networks. If a connection is required between studios on different continents, the choice of service provider becomes important since different providers may use different mechanisms to transmit the data.

This chapter defines telecommunications networks as those parts of data transmission outside of the line of demarcation. In more detail, a telecommunications network is defined as interconnecting a number of stations using telecommunication facilities. The physical circuit between two points is referred to as a link, as

241

opposed to interface, used so far. Confusingly, a node is no longer a single device but a point of junction similar in function to a hub.

8.1.3 Circuit switching

When considering packet switching networks as discussed in Chapter 3, data are expected to be in some message format containing destination addresses. Messages are divided into packets and transmitted over a network of cables interconnected by decision-making devices, which direct the data to their destination.

8.2 Public and private telecommunications networks

There are many telecommunications networks in operation, providing a wide variety of services. A list of these would certainly include public switched telephone network (PSTN), private-line voice networks (PVN), private-line data networks (PDN), packet switching networks, and public switched data networks (PSDN). Services available include audio programme networks such as cable radio, as well as video programme networks, such as cable television and video-on-demand.

The two categories of interest are public switched networks and private- or leased-line networks. The first of these classifications is variously known as public, switched, public switched, message toll service, long distance, direct distance dialling, and inter-exchange facilities. The following terms are used to describe the second of these main categories: private line, leased line, dedicated line, full-time circuit, and tie line.

8.2.1 Public switched networks

Public switched networks provide business and residential telephone services for voice and data transmission to the general public. Users share common switching equipment and channels, and callers wait their turn for service if all the facilities are in use. Fees are typically paid for the use of the network based on the time the call takes, and how much distance it covers, and is very much in the spirit of the traditional commercial telephone model.

Several approaches to the transmission of data over public networks have been taken, including the installation of packet switching public networks to provide a more efficient method of transferring data over networks, as well as the digitizing of voice data to send over private networks.

8.2.2 Private networks

Private networks fall into two distinct groups. Those that are truly private, and those that are leased from, or at least managed by, a telecommunications provider for sole and exclusive use by the customer. A truly private network means responsibility for the entire 7-layer model associations including the transmission media such as cables. Due to the economics, truly private networks on the global scale are relatively rare.

Leased services are more readily available to enterprises needing general communication with a limited number of other locations, or with the Internet. With POTS-type arrangements, services are implemented within the telecommunications provider as a foreign exchange service (FX), implying a telephone exchange 'foreign' to the exchange area where the user is located. A private leased line connected to subscriber premises implements this service to a central office in another exchange. The subscriber can then make an unlimited number of calls to any number within the foreign exchange area for the cost of a local call.

Within data networks, the subscriber normally leases a single point to point connection, of a fixed QoS, calculated on a bandwidth per mile basis. Although these arrangements appear to offer point to point connection, the bandwidth is assigned within the exchange, and a permanent connection is opened through the switches.

8.2.3 Virtual private networks

One valid definition of virtual private networks (VPN) is the use of virtual IP subnets, where a workgroup of computer stations has the same IP subnet address, although data are transmitted on different physical network cables, passing data through a router if necessary.

In the context of telecommunications, a VPN is a mixture of public and private network circuits in a customized arrangement, implying some security for the connection. A leased line from a telecommunications company includes a bandwidth assignment made between named points and assigned at the exchange. Security can be applied using some encryption method.

Virtual private networks are favoured in franchise arrangements, where a cheap/fast general-purpose internetwork connection can be secured over public lines. The central office cannot be dialled into or accessed via the Internet (subject to the exact specification) and so the perception is that the link is free from hacking.

8.2.4 Managing the service

From the point of view of an audio recording studio, as opposed to a broadcast service provider, the responsibility for data transmission once it is outside of their premises, belongs to the telecommunications companies, according to some QoS agreement. Arrangements can be made to lease or rent lines depending upon the national circumstances, but the connection over international boundaries will often be managed by different third parties along the route.

In order to transmit audio over the various connection packages available, it is necessary to understand what these services offer in terms of bandwidth, compatibility, and delay, the key indicators of the quality of the service. Comparison of the figures within the service agreement, against the figures generated by work which the link will be put to, will then give a realistic assessment of whether the proposed investment will meet the requirement.

8.3 ISDN

8.3.1 Introduction

Integrated services digital network (ISDN) is an international concept whose objective is a digital, public telephone network. The implementation is a digital data service capable of transporting any type of data using technology essentially similar to regular telephone calls. The architecture to support this objective builds upon the old POTS infrastructure, although the installation of compliant cables is sometimes required. ISDN is capable of offering digital services to a wide range of consumers and the link allows data to be transmitted using end-to-end digital connectivity.

The form of ISDN of most interest to audio professionals is basic rate interface (BRI), as being the cheapest and most realistic publicly available service. The primary rate interface is the same service anyway, but multiplied by 30 times or so. This is clarified later.

Subscribing

To access the ISDN service, it is necessary to enter a subscription arrangement to an ISDN line. The premises to which the service is to be installed must be within 5.5 km of the telephone company central office for the service to operate correctly. If the service is to be installed in premises outside of this limit, repeater devices are required, or the ISDN service may not be available at all. It should be noted that the distances vary depending upon the international variation in ISDN specifications.

Also required at the customers' premises is the equipment to communicate with the telecommunications provider switch, and with other ISDN devices. These devices include ISDN terminal adapters (sometimes referred to as ISDN modems) and ISDN routers. For audio purposes, a coder–decoder (codec) is generally purchased which formats the incoming signal for ISDN.

There are some differences between ISDN standards in different countries, and this has been one cause of slow consumer uptake of the service.

Provision

There are two basic types of ISDN service: basic rate interface (BRI) and primary rate interface (PRI). BRI is supplied at the line of demarcation in the customer's premises, in the same fashion as a traditional telephone point, but where an adapter must be installed, usually by the provider. It consists of three channels: two channels are used for actual voice or data traffic with each one operating at a rate of 64 KB/s. These are called bearer channels, or B-channels for short. The third channel is used for call supervision and management, performing such tasks as connection and disconnection. This channel operates at a rate of 16 KB/s and is called the delta channel, or D-channel for short. The arrangement of three channels in this manner is referred to as 2B + D and yields an aggregated data rate of 144 KB/s. It is possible to add multiple BRI devices (up to eight) to the same line using the S-bus interface of basic rate ISDN (Telecoms Corner, 2000).

Variations are apparent in the international implementations of the PRI channel structure, amongst others, where the North American implementation has a PRI channel structure of 23 B-channels plus one 64 Kbits/s D-channel, giving an aggregated data rate of 1536 Kbits/s. In most locations, however, it is possible to enter an entirely customized arrangement for whatever bandwidth is required.

In Europe, on the other hand, PRI consists of 30 B-channels plus one 64 Kbits/s D-channel for a total of 1984 Kbits/s. It is also possible to support multiple PRI lines with one 64 Kbits/s D-channel.

H-channels provide a way to aggregate B-channels. They are implemented as:

- H0 = 384 Kbits/s (6 B-channels)
- H10 = 1472 Kbits/s (23 B-channels)
- H11 = 1536 Kbits/s (24 B-channels)
- H12 = 1920 Kbits/s (30 B-channels) – International (E1) only.

Note that, in ISDN terminology, 'K' means 1000 (10^3), rather than 1024 (2^{10}) as mentioned in Chapter 1; therefore a 64 Kbits/s channel carries data at a rate of 64 000 bits per second. Where confusion arises, the reader is encouraged to clarify.

Acceptance and administration

Although available for some time, telecommunications companies have been slow to implement ISDN, especially at the local level. One reason for this is the different practical implementations of the original recommendations, which proved to be incompatible between different vendor networks.

The CCITT began the process of standardization of digital telecommunications and ISDN is documented in CCITT Recommendation I.120 (1984) which describes some initial guidelines. CCITT is now known as the International Telecommunications Union or ITU, and is a United Nations organization responsible for co-ordinating the standardization of international telecommunications (International Telecommunications Union – see Notes and further reading).

During the early 1990s, an industry initiative to establish a more specific implementation for ISDN in the United States began. Members of the industry agreed to create National ISDN 1 (NI-1) so that customers would not have to know the specific implementation of the standard within the switch they are connected to in order to install compatible connection equipment. Once the NI-1 agreement was finalized, a more comprehensive standardization initiative, National ISDN 2 (NI-2), was undertaken. Some manufacturers of ISDN communications, such as Motorola (see Notes and further reading) and US Robotics (http://www.usr.com/) have worked to develop configuration standards for their own equipment.

ISDN allows multiple digital channels to be operated simultaneously through the same regular phone wiring used for analogue lines, provided that the lines in question meet a certain criteria. The change between analogue and digital transmission on a particular line comes about when the telephone company's switches are made to support digital connections. Although the same cable is used, it is a digital signal instead of an analogue signal which is transmitted down the line. This scheme permits a much higher data rate than analogue transmission using traditional modem technology to modulate a signal onto the frequency band used by voice on the telephone cable. In addition, the amount of time it takes to set up an end-to-end conversation between two devices over ISDN is about half that of an analogue link. This has applications for services requiring sustained circuits infrequently, such as interactive applications such as games, video and longer file transmission periods that might be experienced during audio transfer between studios.

The advantage of ISDN when applied to a normal telephone connection is the availability of additional digital services, such as caller ID and multiple lines, with the ability to route calls to different destinations in different ways.

The digital signal also contains information on the caller's number and the type of call (for instance, data, voice, or fax) as well as other enhancements.

Customer equipment

The telecommunications provider supplies BRI customers with a two-wire link from the telephone switch known as a U interface, covered in more detail in the section ISDN line types. The connected device is called a network termination 1 (NT1), which converts the two-wire U interface into a four-wire interface called the send–transmit (S/T) interface. Singularly, S and T are electrically equivalent to each other and the S/T interface can support up to seven devices on the bus. The NT1 device may be in the form of a network interface, allowing support for Ethernet connection, for instance.

ISDN devices pass through a network termination 2 (NT2) device, to convert the T interface into an S interface, and most NT1 devices include NT2 in their design. The NT2 communicates with terminal equipment, and handles the data link and network layer protocols. A basic rate installation of 2B + D is known as ISDN2, as two bearer lines of 64 Kbits/s are supplied.

Devices designed to interface with ISDN are known as terminal equipment 1 (TE1) devices, and have an ordinary telephone interface (known as the R interface).

8.3.2 Narrowband ISDN

Narrowband ISDN offers transport data rates in the order of 1.544 MB/s or less. This capacity has been configured in several general service offerings discussed below.

Circuit switched voice

This is a digital voice service offered over a four-wire ISDN digital subscriber line (DSL), with customer equipment normally located inside the premises.

Circuit switched data

Circuit switched data services provide point to point digital data connection over the public network (see VPN). ISDN uses separate signalling channels for the establishment and maintenance of data connections requiring special processing (see ATM). The customer equipment will normally be purchased as part of the cost of installation and may be retained by the customer. Service agreements cover responsibility thereafter.

Low speed packet

A monitoring capability is provided by using the D-channel on DSL. The D-channel provides a data rate in the region of 16 KB/s using the X.25 protocol within the upper layers. The D-channel can be used for low speed packet data, whilst also relaying call-processing information. This can be aggregated with the high-speed packet service to obtain the maximum data rate.

High speed packet

High speed packet service describes the configuration of one or both of the B-channels in 64 Kbits/s partitions used for circuit switched voice, circuit switched data, or high speed packet service in a user definable combination.

8.3.3 Broadband ISDN service

Broadband ISDN service can supply data rates beyond the 1.544 MB/s limit of narrowband, and is usually run on a frame relay, SMDS, or ATM connection-based network. The data rates available for broadband ISDN services range from 25 MB/s up to gigabits per second, and are implemented within backbones.

SONET

Synchronous optical networks (SONET) are based on fibre media for physical layer transmission. As a physical transmission mechanism, standardized SONET technology supports flexible transport architectures. The two most common speeds of broadband ISDN running over SONET type networks are 155 MB/s (OC-1) and at 622 MB/s (OC-3), made possible by the high quality of the digital facilities in place on the network.

Frame relay

Frame relay supports the transport of data in a connectionless service, meaning that each data packet passing through the network contains address information. Common frame relay services start at 56 Kbits/s and 1.544 Mbits/s, scaling to 25 MB/s and beyond. One unique facet of frame relay service is the ability to support variable size data packets. This has applications for variable bit rate (VBR) data flows, such as might be utilized within interactive applications and simulation.

Switched multimegabit digital service

Switched multimegabit digital service (SMDS) is a digital service providing a high speed digital path for permanent virtual circuits (see VPN). Common service offerings provide data rates from 56 Kbits/s to 34 Mbits/s and up to 155 Mbits/s and beyond, making it scalable from the customer's perspective. As with SONET, the link is transparent to upper layer protocols, meaning that the functionality of the upper layers can be accommodated. So, for instance, using the TCP/IP protocol suite over SMDS allows virtual subnetworks to be implemented.

8.3.4 Physical layer

The ISDN physical layer is specified by recommendations published in the ITU I-series and G-series of documents (http://www.itu.ch/publications/itu-t/itutg.htm). Around 800 documents are contained within the G-series and so the full list is not reproduced here! The documents are available at the ITU website, where each descriptive title links to a download of the full text. As an example of the wide scope of the recommendations, one section of the list is entitled 'General recommendations on the transmission quality for an entire international telephone connection' and this section contains recommendation G.111 (3/93) loudness ratings (LRs) in an international connection, amongst others.

Local loop and backbone

The connection between the customer equipment at the line of demarcation and the telephone switch is called the local loop. Because of the many miles of cables and the variety of physical locations required to service millions of telephone outlets, the local loop is the most difficult and expensive to install and upgrade.

The local loop connection is called the line termination (LT function), whilst the connection between switches within the phone network is called exchange termination (ET) function. The interface between the local loop and the main telecommunications backbone is known as the V interface.

ISDN line types

ISDN installation depends upon the characteristics of the local loop. Before the service provider installs an ISDN line, a series of measurements are taken to ensure that the cable pairs are loop qualified and meet the requirements of ISDN. If the current analogue line does not meet the requirements, a new cable is required.

T-interface

All ISDN lines are equipped with an individual line card in the central telecommunications switch, located at the provider's premises. ISDN users within approximately 1000 m of the local ISDN switch may be provisioned on a T-interface line. A T-interface line provides a direct ISDN connection to the switch and is a four-wire link. A T-interface line delivers the ISDN signal to the terminal since this is equivalent to having the NT1 equipment housed at the provider's convenience, although the line is actually connected to a card in the switch.

U-interface

The U-interface supports a longer local loop than the T interface. The U-interface line uses a different type of interface card at the switch, and is supplied as a two-wire interface. The length restriction depends upon the encoding technique and wire gauge (on the 2B1Q encoding technique, explained later, the limit is 9000 m of 33 gauge wire or 42 dB loss at 40 kHz). The U-interface also needs to have an NT1 unit located at the user's premises.

Z-lines

The Z-card line is used where an analogue line may be required in a location served by predominantly ISDN lines. Z-cards are most commonly used for telephones in environmentally difficult areas or for pay telephones.

ISDN terminals

Power

All ISDN terminals require a power connection and so ISDN is not available in the event of a power failure. The power is used for the terminal encoding and decoding and to provide signalling tones. Unlike an analogue telephone wherein a ringing generator signal is placed on the line to ring the phone, in ISDN an

'Alerting' call control message is relayed to the terminal. The 'Alerting' message tells the ISDN terminal to play certain tones. The dial tone originates within the ISDN terminal.

Encoding techniques

As mentioned earlier, the difference between ISDN standards across national boundaries causes incompatibility when interconnecting between implementations. In the physical layer, an example of this can be found within the encoding mechanisms for representing binary states as voltages on the cable. BRI and PRI are different even in the same country, because of the different work that each link type performs, but there are also differences between BRI services between countries.

In America, the BRI service uses the 2B1Q scheme to represent data as voltages on the cable. Europe, on the other hand, uses a technique called 4B3T for BRI services. Both schemes are used in mechanisms achieving higher data rates than are currently available for LANs.

2B1Q

2B1Q is detailed in the 1988 ANSI specification T1.601 which describes a scheme whereby the input voltage level can be one of four distinct levels. As illustrated in Figure 8.2, a level can represent two binary states, since there are 2^2 possible variations. Note that a voltage level of zero is not valid under this scheme. The figure shows the voltage levels and the pairs of bit states that each represents.

Figure 8.2 2-Binary 1-Quaternary (2B1Q) encoding scheme, illustrating the concept of representing more than one bit state with each possible voltage state. Using 2B1Q, a possible four states exist on the media, and each state represents two bit positions.

Bits	Voltage
00	-2.5
01	-0.833
10	+2.5
11	+0.833

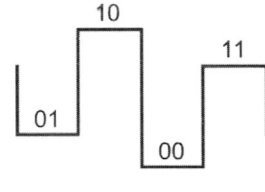

4B3T

4B3T uses return-to-zero (rather than NRZ) encoding to represent 4 bits rather than 2 bits, within one voltage state. Because of this relation to one voltage state equalling 4 bits, the voltage state is called a quaternary.

4B3T is defined in European Telecommunications Standards Institute (ETSI) ETR 080, Annex B and other national standards, such as Germany's 1TR220. 4B3T can be transmitted reliably at up to 4.2 km over 0.4-mm cable, or up to 8.2 km over 0.6-mm cable.

PRI encoding varies depending upon use, and research is continuing to improve encoding techniques (Telecoms Corner, 2000). In North American PRI, transmission of T1 grade service normally occurs over twisted-pair cable, using one of two types of line coding – AMI or B8ZS. Line coding is defined in ITU recommendation G.703 and a variety of US standards.

AMI line coding

Alternate mark inversion (AMI) is used in North American T1 (1.544 MB/s) implementation. With AMI, a positive or negative pulse is generated with each mark or 1. The pulse polarity depends upon the polarity of the last preceding pulse. When a space or 0 is transmitted, there are no pulses generated. Repeaters usually require pulse transitions to recover and regenerate timing on the line, requiring that long strings of 0s be avoided. The synchronization method employed within ADAT uses a similar rule but AMI specifies that no more than 15 consecutive zeros are allowed.

B8ZS line coding

Bipolar 8 zero substitution (B8ZS) is a newer line-encoding scheme used within North America. AMI timing recovery problems encountered with consecutive zeros, is solved with B8ZS coding. One of two special bipolar violation codes is substituted for strings of eight zeros. Intentional bipolar violations are caused in bit positions 4 and 7 of the data stream. With B8ZS encoding, no more than seven consecutive zero voltage states will be transmitted on the line. ADAT also uses a deliberate violation pattern in order to facilitate frequency selection.

The violation codes cause two bipolar violations of opposite polarity in the bit stream, which also thus have the effect of preventing excessive DC voltage levels.

8.3.5 Data link layer

The ISDN data link layer is specified by the ITU Q-series documents Q.920–Q.923. Apart from the encoding previously described, the remainder of the layer concerns itself with synchronization and error checking.

Frame format

A US T1 frame consists of 193 bits: 192 bits contain 1 byte of data from each of the 23 B-channels and 1 byte of data from the

D-channel (8 bits $\times$ 24 bytes = 192 bits). A single bit is prefixed to the data as a stop-bit (or point of reference) for the synchronization.

Frames can be combined into a super-frame or an extended super-frame, which are made up of 12 and 24 frames, respectively.

A European E1 frame consists of 256 bits: 240 of these are used to transport 1 byte of data for each of the 30 B-channels plus the D-channel and 1 byte for framing and synchronization purposes. Sixteen frames are combined to form a multi-frame.

Synchronization

The synchronization pattern within the BRI frame contains the nine quaternaries DC pattern, followed by 12 occurrences of data, aggregated in the form $B_1 + B_2 + D$. Each aggregation is 18 bits in length, made up of 8 bits from each channel. Finally, the frame is completed with a maintenance field, consisting of redundancy checks, error flags, and sysex commands used for loop-back testing without disrupting user data.

When bundled together in a super-frame, the sync field of the first frame is made significant by inversion of the DC phase, and this represents a string of six consecutive 1s (or 7E h). 7E h is known as the flag character and is part of the header information.

8.3.6 Network layer

The ISDN network layer is also specified by the ITU Q-series documents Q.930–Q.939.

Addressing

Basic addressing occurs in the network layer, and this is contained within the call reference field. The call reference field consists of 2 or 3 bytes. BRI systems use a 7-bit call reference value (with 128 possible combinations) within 2 bytes. PRI systems have a 15-bit call reference value (yielding 32767 possible combinations).

The call reference value is an arbitrary number, which represent a conversation, or logical link, between the two end devices, allowing devices to manage multiple conversations on the same cable.

Figure 8.3 shows a cell-like network of routers. The value in the call reference field between points A and B will not necessarily

Figure 8.3 Addressing in
ISDN uses a call reference
number to identify a
connection within the switch.
This is not related to any
other network addressing that
might occur within the data,
such as IP addresses.

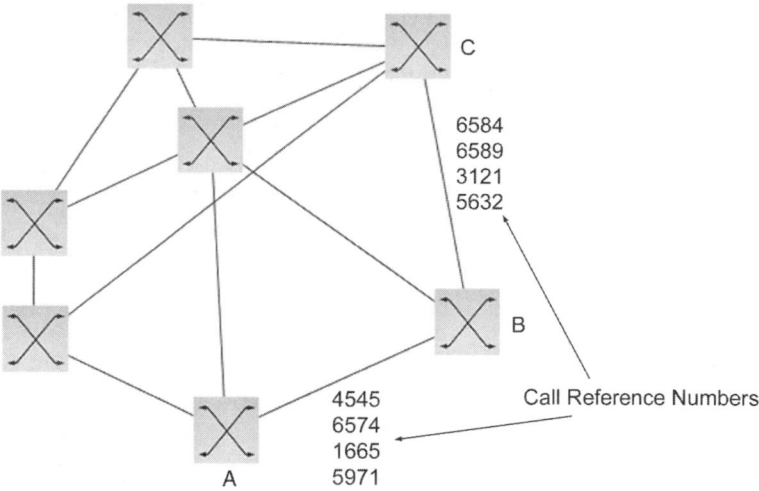

6584
6589
3121
5632

Call Reference Numbers

4545
6574
1665
5971

A

B

C

be the same as between points B and C, but serves as an
index number that both nodes can refer to when managing
connections.

Message types

There are four general message types: call establishment, call
information, call clearing, and miscellaneous. Obvious candidates for inspection are call establishment and call clearing
messages. The full list of bit pattern assignments is shown in
Figure 8.4. The message type byte identifies the exact message
and thereby determines what additional information is required
and allowed.

Call set-up

In order for a communications channel to be established, the
initiating device first sends out a set-up (00101) signal across the
D-channel. Under normal circumstances, the switch at the
opposite end of the cable, labelled as point B in Figure 8.3, sends
a call proceeding signal (00010) back to the initiator, as well as a
set-up signal to the intended recipient.

Upon receipt of the set-up message, the recipient device (C)
sounds the ring of the telephone and sends an alert signal (00001)
back to the switch. This is then forwarded to the initiator (A).
When the call is answered, device C sends a connect message
(00111) to the switch which is then forwarded to device A. An
acknowledgement message (01101) is then returned and the call
can proceed.

Figure 8.4 Message types in ISDN.

0000	**Call Establishment**	
	00001	Alerting
	00010	Call Proceeding
	00111	Connect
	01111	Connect Acknowledgement
	00011	Progress
	00101	Set up
	01101	Set up Ack
001	**Call Information**	
	00110	Resume
	01110	Resume Ack
	00010	Resume REJ
	00101	Suspend
	01101	Suspend Ack
	00001	Suspend REJ
	00000	User Information
010	**Call Clearing**	
	00101	Disconnect
	01101	Release
	11010	Release Complete
	00110	Restart
	01110	Restart Ack
011	**Miscellaneous**	
	00000	Segment
	11001	Congestion Control
	11011	Information
	00010	Facility
	01110	Notify
	11101	Status
	10101	Status Enquiry

8.3.7 Audio implementations

The broadcast industry is able to make use of equipment, allowing the ability to talk off-air with a caller at the same time as other callers are on-air. Such equipment might also include some DSP before the compressing the audio, such as through MPEG Layer III, for better data density. Although some audio artefacts may be lost during such compression, the improvement in transmission efficiency is considered appropriate. Compression technologies are important to the study of audio transfer, although such is the complexity of the subject that further study is recommended!

MPEG Layer III encoding is used to offer full-duplex 15 kHz stereo audio or simplex streamed audio transmission, for which specialist equipment must be purchased.

Applications for this technology might include Internet audio, remote broadcasting, video conferencing, or logging.

8.4 Presto

Presto is a data link interface designed to map audio, video, MIDI, Ethernet, and synchronization data onto the ANSI T1.105–1995 synchronous optical network (SONET). Presto can be implemented as a local interface on a PC board, a point to point interface, a local area network, or a wide area network. Presto can transmit more than a thousand audio channels over fibre optic cable for distances up to 80 km. Presto supports ANSI SONET STS-3c/OC-3 (155 Mbits/s), STS-12/OC-12 (622 Mbits/s), and STS 48/OC-48 (2.49 Gbits/s). It can also be transmitted at OC-3 speeds using category 5 cable among other media types.

There is room for expansion with fibre optic cable lengths of 40 km and longer and speeds of 9.95 Gbits per second and greater.

8.4.1 Standards and administration

Presto is an initiative undertaken by Kurzweil Music Corporation with revision 0.4 of the specification made available shortly after its publication date of 4 September 1999 (*Professional Audio via Synchronous Optical Network* – see Notes and further reading).

8.4.2 Timing considerations

Like SONET, Presto is synchronous in nature. Analogue conversion clocks are synchronized to the interface clock rate or vice versa. Various audio sample rates are supported, from 22.05, 24, 32, 44.1, 48, 64, 88.2, 96, 128, 176.4, 192, 256 kHz, 2.8224 and 5.6448 MHz, and sample rates with rational fraction relationships to the 8 kHz SONET frame rate. These rational fractions are used to synthesize conversion clocks from a clock source. The clock is manipulated using a 24-bit number, allowing fine adjustment of sample clock frequency. Sample frequency can also be adjusted without requiring additional buffering, or other causes of delay. Sample word widths of 1, 16, 24, and 32 bits are supported. The maximum capacity of audio-only transport ranges from 128 channels at 48 kHz, 24 bits using STS-3c to 2048 channels at 48 kHz, 24 bits using STS-48.

8.4.3 Data format

In order to reduce delays and buffer requirements when interfacing with ADAT, TDIF, and other audio interfaces, data are transmitted most significant bit first, and multiplexed as pairs of channels (a/b) or as single channels. Bit reversal within samples is required when interfacing directly to AES3. The most significant data byte within a channel is sent first and is numbered byte 1. The next most significant byte sent next is numbered byte 2, and so on.

Audio is multiplexed by channel, with the most significant byte of the 1st channel transmitted first, the most significant byte of the 2nd channel second, and so on. Once the last byte of the last channel in the last set has been transmitted, the payload is padded.

8.4.4 Error correction

To decrease the chance of an interface error causing an audible malfunction, data can be transmitted in duplicate. If a CRC error indicates that one copy of the data may be bad, the other copy can be used instead. To accomplish this, at rates of 96 kHz and below, data are transmitted in pairs at twice the normal rate, with channel pair sets carrying a single channel of data each. Once all bytes from all channels have been transmitted once, they are transmitted again. There is at least one 16-bit CRC for each copy of data transmitted.

8.5 Asymmetrical techniques

Asymmetric techniques allow higher data rate in one direction than in the other. From a subscriber's point of view, this means that lots of data can be received, but fewer data can be sent out in the same time period. Asymmetrical techniques are useful in instances such as video-on-demand, where the data rate coming into a subscriber's premises is required to be much higher than the outgoing data rate. An imagined video-on-demand service, for instance, might require that a subscriber make a request to view a video at a pre-selected time. The time that the video is transmitted to the subscriber's premises might be one of several predetermined times made available by the video provider, or transmitted at the subscriber's request, beginning seconds after the request is made, as described in Chapter 1. In either case, the data sent out from the subscriber whilst making the booking are only a tiny fraction of the data received whilst viewing the video.

Another service, to which asymmetrical techniques are being applied, is fast Internet access. This is possible because of the behaviour of the vast majority of users of the Internet, where a relatively small address is sent as a request for a larger amount of information.

8.5.1 ADSL

Asymmetric digital subscriber line (ADSL) is a subscription-based service used to deliver higher rates than BRI ISDN over the existing POTS infrastructure.

ADSL is one member of a family of transport systems called digital subscriber line (xDSL) and practical implementations are capable of offering data rates between 1 and 10 MB/s over copper lines. Outgoing data rates from the subscriber's premises are in the region of 64–640 Kbits/s.

ADSL was motivated by a requirement to make available the use of a regular telephone connection in the event of a power failure, offering an improvement in service over BRI ISDN. In general, the fastest DSL interfaces can only be supported within 2 km of the switch, although the implementations offering slower data rates can go further.

General description

Subscription to an ADSL circuit requires an ADSL modem on each end of a twisted pair telephone line. Three communication channels are created on the cable, and these are split into a high speed downstream channel, a medium speed duplex channel that depends on the implementation of the ADSL architecture, and a POTS or ISDN channel. The POTS/ISDN channel is removed from the digital modem by filters allowing uninterrupted POTS/ISDN, in the case that the ADSL portion fails. The high speed channel ranges from 1.5 to 8 Mbits/s, whilst the duplex rate ranges from 16 to 640 Kbits/s. Each channel can be subdivided to form multiple, lower rate channels.

Dividing the available bandwidth of a telephone line in one of two ways creates the multiple channels. Frequency division multiplexing (FDM) is used to assign one band for upstream data and another band for downstream data. The downstream path is then further divided into one or more high speed channels and one or more low speed channels. The upstream path is also multiplexed into low speed channels. With either technique, ADSL splits off a 4 kHz region for POTS.

The implementation of FDM is of interest because it is this that allows ADSL to be delivered through the various quality cables that make up the POTS infrastructure. In more detail, frequencies are divided into 4 kHz blocks, each one becoming a data channel in its own right, with POTS making up a specific frequency range. Where poor quality cable impedes the transmission of a certain channel, ADSL has the ability to cease transmission on that channel, and adjust and manage the data rate accordingly. Some configurations of the POTS infrastructure, such as the presence of line amplifiers with certain electrical properties on the link, can completely obstruct all channels, meaning that ADSL cannot be usefully used to transmit data. Customer experience of ADSL may vary, depending on factors including the distance from the provider's premises and the quality of the available POTS line.

Another advance which went into ADSL was in the error correction mechanism. As a real-time signal, digital video cannot use link- or network-layer error control, for reasons previously explored. However, ADSL incorporates a forward error correction mechanism which reduces errors caused by impulse noise, thereby removing a significant cause of errors. This technique introduces a delay in the order of 2 ms.

The data format within a channel is again frame based, with an error correction code attached to each frame or block. The receiving modem corrects errors that occur during transmission up to the limits implied by the code and the block length. Optionally, the unit may also create super-blocks by interleaving data within sub-blocks, thereby allowing the receiver to correct any combination of errors within a specific span of bits. Typically, an error rate of 1 in 10 000 000 bits or higher is achieved, allowing for effective transmission of both data and video signals. Video compression technology allows transmission requiring a data rate of around 1 MB/s for VCR quality and 2–3 MB/s for broadcast quality pictures (Direct TV satellite systems use approximately 3–4 MB/s. Source: ADSL Forum, see Satellite). Practical implementations have achieved data rates beyond the originally anticipated 6 Mbits/s and led to the transmission of at least one such video channel or a number of audio channels (ADSL Forum, 1998).

Because of the wide variety of environments characterizing the local loop, problems occur in the implementation of digital technology. Long telephone lines may attenuate signals at 1 MHz (the outer edge of the frequency range band used by ADSL) by up to 90 dB. Developments in frequency splitter technology have allowed the realization of a large dynamic range, with low noise figures.

ADSL can be purchased with various speed ranges and capabilities. The minimum configuration provides 1.5 or 2.0 Mbits/s downstream and a 16 Kbits/s duplex channel. Improvements continue to increase the upper data rates available to subscribers.

Standards and associations

Although dispute regarding line coding techniques (*Discrete Multitone* . . . – see Notes and further reading) prevented early adoption, ANSI administrates the standard and technical recommendations within the North American continent. Within Europe, administration is performed by the European Technical Standards Institute (ETSI – see Notes and further reading).

The ANSI working group T1E1.4 approved an ADSL standard for rates up to 6.1 MB/s (ANSI Standard T1.413) (Alliance for Telecommunications Industry Solutions – see Notes and further reading). ETSI contributed an Annex to T1.413 to reflect European requirements. T1.413 currently embodies a single terminal interface at the subscriber's premises. Subsequent issues include a multiplexed interface with additional recommendations for configuration, network management, and so on. The ATM Forum and the Digital Audio Visual Council (DAVIC – see Notes and further reading) recognize ADSL as a physical layer transmission protocol for unshielded twisted pair media.

DAVIC was created in August 1994 with an expected duration of 5 years. Its task was to create complete sets of specifications using emerging digital audio–visual technologies. The term of 5 years was reached in August 1999 and the set of specifications (DAVIC1.3.1) had become an International Standard and International Report (IS 16500 and IR 16501) after the resolution of comments performed in June 1999.

The ADSL Forum develops technical guidelines for architectures, interfaces, and protocols for telecommunications networks incorporating ADSL transceivers. The ADSL Forum was formed in December 1994 to represent commercial organizations involved with ADSL and to lend the weight of practical implementations to such areas as system architectures, protocols, and interfaces. In November 1999, the Forum voted to change the name, by dropping the A. It felt that the adoption of the new name would demonstrate more clearly that the Forum is an all-embracing worldwide body, covering all the varieties of DSL, and the name was subsequently changed in January 2000 (DSL Forum – see Notes and further reading).

Physical layer

Cable

As mentioned, the performance or availability of ADSL to particular premises relies on the proximity to the telecommunications facility, or in other words, the length of the local loop. The

Figure 8.5 Table showing data rate/distance ratio for ADSL performance expectations.

Data Rate	Wire Size	Distance
1.5 or 2 MB/s	0.5 mm	5.5 km
1.5 or 2 MB/s	0.4 mm	4.6 km
6.1 MB/s	0.5 mm	3.7 km
6.1 MB/s	0.4 mm	2.7 km

table in Figure 8.5 indicates the data rate/distance ratio and the types of cable expected. The cable type is twisted pair.

ANSI T1.413 specifies the frequency range of 26 kHz to 1.1 MHz, with frequencies below 4 kHz reserved for POTS. This means that the use of a normal telephone over an ADSL connection will not affect the data rate of the connection.

Load coils used on voice lines to improve the quality of the voice service do not allow signals above the voice band 0–4 kHz to pass through them, meaning that ADSL is not available on such lines.

Encoding

Implementations of DMT within ADSL divide the downstream channels into 256 4 kHz wide tones and the upstream channels into 32 sub-channels. Data are allocated onto a sub-channel and one sub-channel transmits a stream of data serially. In this way, a parallel transmission is achieved. Transmission of data is turned off, for instance if a particular sub-channel is prone to RF interference over a particular connection.

Combining ISDN and ADSL signals

The methods for combining ISDN and ADSL (*Transmission and multiplexing* – see Notes and further reading) can be placed into the general categories of in-band and out-of-band, and these are illustrated in Figure 8.6.

In-band combinations treat the ISDN signal as another stream of data to be carried by ADSL. This approach has the advantage that the ADSL modem is free to use the standard frequencies for transmission and start-up. Using this method allows compliance with the T1.413 standard, but means that the ISDN signal, including the voice component, is processed by the ADSL modem. Since the modem requires local power, a local power failure will also cause the lifeline telephone service to fail. Furthermore, the processing delay of around 2 ms, introduced by the forward error correction to compensate for impulse noise, is more than the maximum delay of 1.25 ms tolerated by ISDN.

Out-of-band combinations leave the ISDN signal intact, to be transmitted separately at its usual frequencies. The ADSL signal turns off channels that encroach upon ISDN bandwidth, thereby

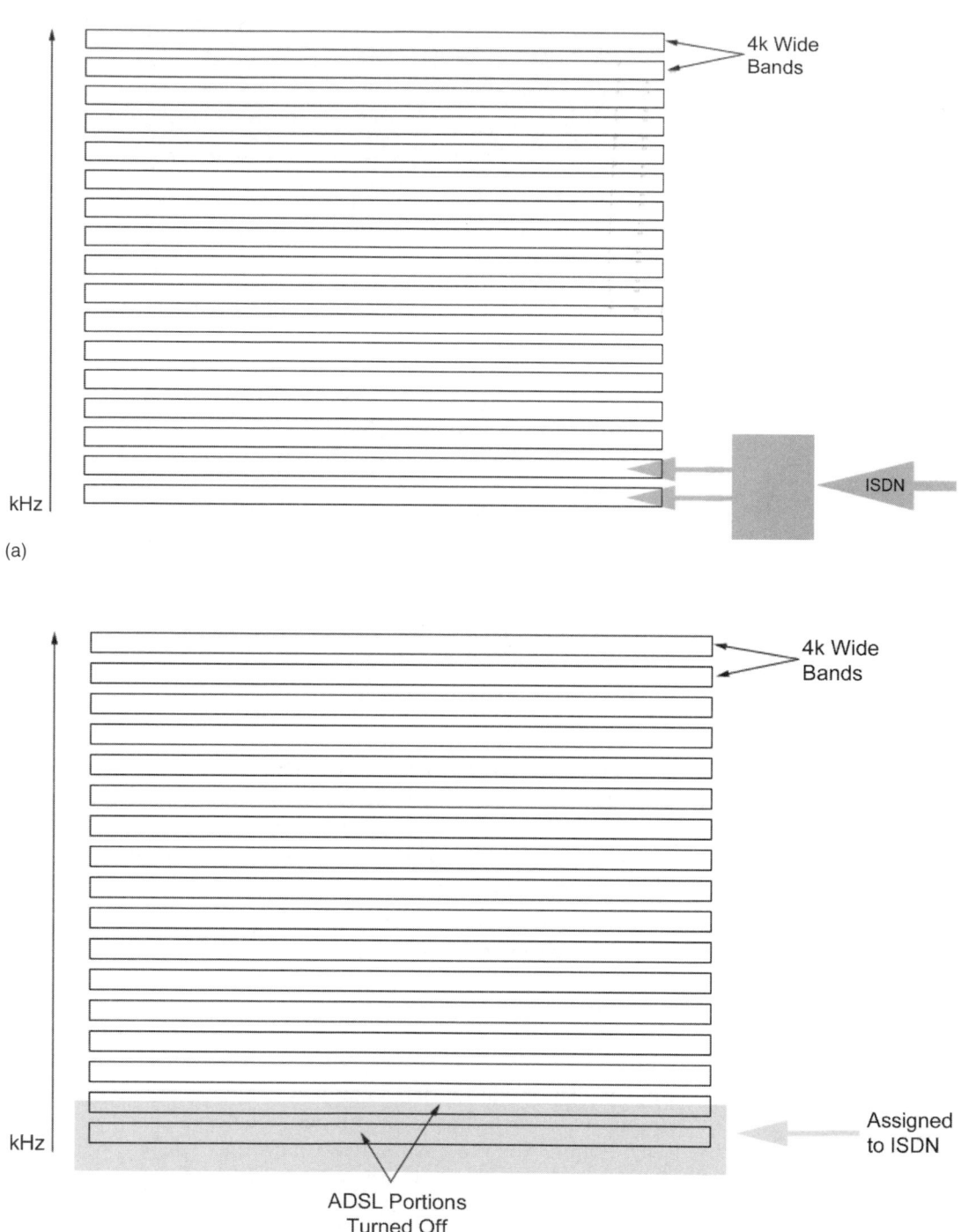

Figure 8.6 Methods for combining ADSL and ISDN. (a) In-band. (b) Out-of-band.

restricting ADSL to higher frequencies. This technique requires a departure from the T1.413 standard, although the benefits of the out-of-band method are substantial and include the lifeline telephone connection, thereby justifying the required deviation from the standard within practical implementations. A passive filter is required to correctly separate the ISDN and ADSL signals from one another (Orckit 38 – see Notes and further reading).

Applications

ADSL has been applied to video-on-demand, and this is often enhanced with additional consumer-oriented applications taking advantage of the available data rate in each direction. Another reason for the interest amongst consumers has been for fast Internet access whilst protecting the investment in POTS. Since ADSL can use large proportions of the copper infrastructure, and no additional cables need to be laid in the majority of areas, the service is a cheap way to increase the inbound data rate.

In practice, the benefit depends upon the distance from the switch, and may be completely impeded by line-driving equipment. In short, customers' experiences have been varied in the early stages of deployment.

A practical application is the DSL-Lite initiative, which envisages a USB interface within the ADSL modem. Administration for DSL-Lite is performed by the UAWG and ITU.

8.6 Satellite communications

The use of satellite communications as a consumer technology really started at the end of the 1990s with the retasking of redundant military satellite hardware on a commercial scale, following the end of the cold war.

From the consumer perspective, transmitting to a satellite is expensive, because of the high powered equipment required for the job, and the financial arrangements with the satellite owners. However, it is relatively cheap to receive information transmitted from the satellite, and a simple example of this can be seen in the prevalence of satellite television such as Direct TV in North America and parts of Europe. Satellite television is very much in the broadcast industry sector and subscription involves a small dish installed at the consumer's premises. In this way, large amounts of data can be received once the dish has been correctly aligned with the satellite. Satellite communications facilities offer various services, with Direct TV able to supply a selection of 200 video channels.

Apart from television, the other use to which this kind of communication has been put is fast Internet access. In order to overcome the practical problems of two-way communication through a satellite, the outgoing data are handled by transmission through a link such as ISDN BRI, or telephone modem, to the operator of a satellite transmission premises. The uplink transmits the data from hundreds or thousands of sources through a large dish or base station. The satellite then bounces the data back down to the ground for receipt by the consumer's dish, thereby completing the loop.

Service offerings are scalable and bandwidth can be supplied on demand or as a commodity. For instance, if a television broadcast is required, bandwidth can be purchased on the link in order to do this, with the bandwidth being reallocated once it is no longer required. Such a system might be adopted for the broadcast of business television.

Figure 8.7 Overcoming line-of-sight difficulties using multiple satellites.

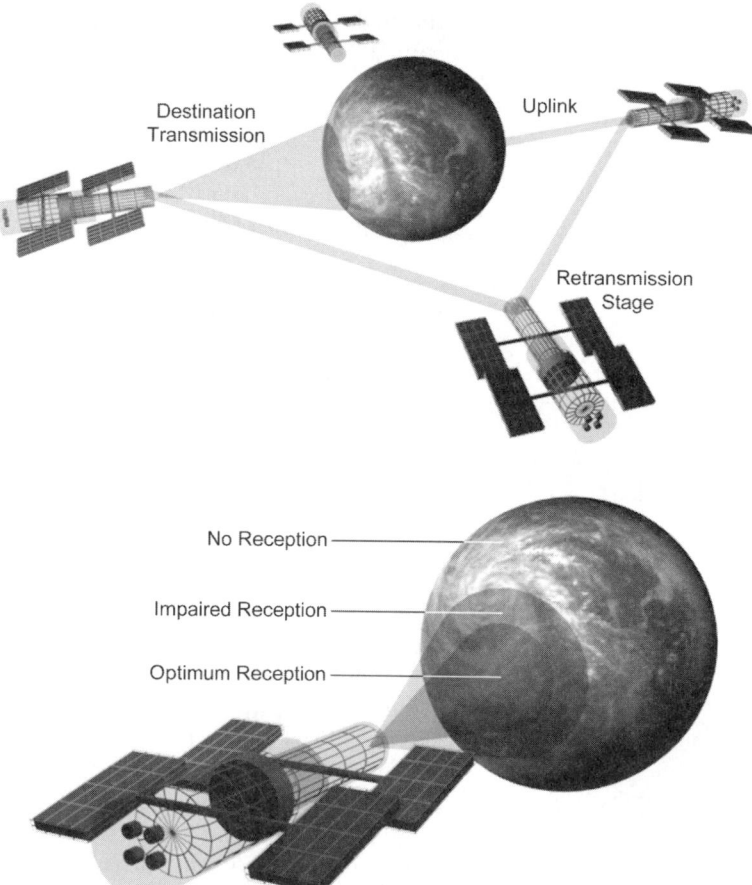

Figure 8.8 Footprint of satellite. Within the circular pattern shown on the globe, good reception can be expected. Reception will be impaired or impossible outside of this area without the use of further facilities.

Satellite communication can be made global by bouncing the signal from satellite to satellite, thereby overcoming line-of-sight problems encountered due to the curvature of the earth (see Figure 8.7).

Shown in Figure 8.8, satellite communication is associated with a footprint, since one geostationary satellite is able to transmit permanently to a certain area, depending upon its location. Orbit types can be classified in a number of ways, such as high and low orbit, and geostationary or geomotionary. High orbit satellites have a longer latency incurred simply because of the distances that the signal must travel to and from the satellite, but high orbit satellites also have a larger footprint. Geostationary satellites move at the same speed as the rotation of the earth and therefore maintain a constant presence in a particular position in relation to the ground. These types of satellite are more generally suitable for the provision of telecommunication because of their static nature.

8.7 IBM network

The IBM network is chosen as an example of a privately owned global network by virtue of one of the first global data networks to be used commercially. The International Business Machines Corporation (IBM) built the network principally for the transmission of the proprietary systems network architecture (SNA) protocol. More recently, the network has been made capable of transmitting IP, with the result that it has become a significant part of the Internet. This is true of only a handful of commercially owned networks.

Connection to the IBM network must be made by subscribing to the appropriate Internet provider, or by purchasing a telecommunications link to the nearest network node. Network nodes are strategically placed internationally, covering many major cities.

On 8 December 1998 AT&T and IBM announced a series of agreements under which AT&T acquired IBM's global network business for $5 billion in cash. The contract for the US network was signed on 30 April 1999, and subsequently in other countries.

Various service offerings include remote access services, IP remote access, DSL and Internet VPN gateway and global managed Internet service (AT&T – see Notes and further reading), such as was used for the IEEE 1394 over long-distance tests, mentioned in Chapter 5.

IBM offers consumer level Internet access over the same network. From the consumers' point of view, globally managed network provision uses the same cables as the Internet access, although the bandwidth is guaranteed for the entire period of the contract, and so it is possible to purchase very high data rates permanently attached across the globe.

Notes and further reading

ADSL Forum (1998) *Frequently Asked Questions.* DSL Forum Office, 39355 California Street, Ste. 307, Fremont, CA 94538, USA. http://www.adsl.com.tech_faqs.htm.

Alliance for Telecommunications Industry Solutions. 1200 G Street, NW, Suite 500, Washington DC 20005, USA. http://www.t1.org.

Appleby, S. (1994) *BT Technology Journal,* April, pp. 19–29. British Telecom Res. Labs, Ipswich, UK.

AT&T. 295 North Maple Avenue, Basking Ridge, NJ 07920, USA. http://www.att.com/globalnetwork.

Digital Audio Visual Council (DAVIC), Geneva, Switzerland. Only available at http://www.davic.org.

Discrete Multitone (DMT) vs. Carrierless Amplitude/Phase (CAP) Line Codes (2000) Aware Technologies, 40 Middlesex Turnpike, Bedford, MA 01730, USA. http://www.aware.com/technology/whitepapers/dmt.html.

DSL Forum. 1212 Suffolk Street, Naperville, IL 60563, USA. http://www.adsl.com.

European Telecommunications Standards Institute (ETSI), 650 Route des Lucioles, F-06921 Sophia Antipolis Cedex, France. http://www.etsi.org.

International Business Machines Corporation. New Orchard Road, Armonk, NY 10504, USA. http://www.ibm.com.

International Telecommunications Union (ITU), Place des Nations CH-1211 Geneva 20, Switzerland. http://www.itu.int/.

Internet2 International. University Corporation for Advanced Internet Development, 1112 16th Street, NW Washington DC 20036, USA. http://www.internet2.org.

Motorola Incorporated. European Headquarters, Church Road, Lowfield Heath, Crawley, West Sussex, RH11 0PQ, UK. http://www.mot.com/.

Orckit 38 Nahalat Yitzhak Street, Tel Aviv 67448.

Professional Audio via Synchronous Optical Network (Presto) (1999) Rev: 0.4. Kurzweil Music.

Telecoms Corner (2000) *Technical Reference site.* http://telecom.tbi.net/.

Transmission and multiplexing (TM); access transmission systems on metallic access cables; asymmetric digital subscriber line (ADSL) – coexistence of ADSL and ISDN–BA on the same pair. ETSI TS 101 388.

Index

Focal Press

http://www.focalpress.com

Join Focal Press On-line

As a member you will enjoy the following benefits:

- an email bulletin with **information on new books**
- a bi-monthly **Focal Press Newsletter**:
  - o featuring a selection of new titles
  - o keeps you informed of **special offers, discounts and freebies**
  - o alerts you to **Focal Press news and events** such as author signings and seminars
- complete access to **free content** and reference material on the focalpress site, such as the focalXtra articles and commentary from our authors
- a **Sneak Preview** of selected titles (sample chapters) *before* they publish
- a chance to have your say on our **discussion boards** and **review books** for other focal readers

Focal Club Members are invited to give us feedback on our products and services. Email: worldmarketing@focalpress.com – we want to hear your views!

Membership is FREE. To join, visit our website and register. If you require any further information regarding the on-line club please contact:

Emma Hales, Promotions Controller
Email: emma.hales@repp.co.uk
Fax: +44 (0)1865 315472
Address: Focal Press, Linacre House,
Jordan Hill, Oxford,
UK, OX2 8DP

Catalogue

For information on all Focal Press titles, we will be happy to send you a free copy of the Focal Press catalogue:

USA
Email: christine.degon@bhusa.com

Europe and rest of World
Email: carol.burgess@repp.co.uk
Tel: +44 (0)1865 314693

Potential authors

If you have an idea for a book, please get in touch:

USA
Terri Jadick, Associate Editor
Email: terri.jadick@bhusa.com
Tel: +1 781 904 2646
Fax: +1 781 904 2640

Europe and rest of World
Christina Donaldson, Editorial Assistant
Email: christina.donaldson@repp.co.uk
Tel: +44 (0)1865 314027
Fax: +44 (0)1865 315472